Hominid Sites: Their Geologic Settings

AAAS Selected Symposia Series

Published by Westview Press, Inc.
5500 Central Avenue, Boulder, Colorado

for the

American Association for the Advancement of Science
1776 Massachusetts Avenue, N.W., Washington, D.C.

Hominid Sites: Their Geologic Settings

Edited by
George Rapp, Jr., and Carl F. Vondra

AAAS Selected Symposium **63**

AAAS Selected Symposia Series

This book is based on a symposium which was held at the 1980 AAAS National
Annual Meeting in San Francisco, California, January 3-8. The symposium was
sponsored by AAAS Sections E (Geology and Geography) and H (Anthropology).

Published in 1981 in the United States of America by
    Westview Press, Inc.
    5500 Central Avenue
    Boulder, Colorado 80301
    Frederick A. Praeger, Publisher

Library of Congress Cataloging in Publication Data
Main entry under title:
Hominid sites: their geologic settings.
    (AAAS selected symposium ; 63)
    "Based on a symposium which was held at the 1980 AAAS national annual
meeting in San Francisco, California, January 3-8"--T.p. verso
    Includes bibliographies.
    1. Fossil man--Congresses. 2. Paleontology, Stratigraphic--Congresses.
3. Excavations (Archaeology)--Congresses. I. Rapp, George, Jr., 1931-
II. Vondra, Carl F. III. American Association for the Advancement of Science.
IV. Series.
GN282.H65   930.1   81-10313 ✓
ISBN 0-86531-262-1   AACR2

Printed and bound in the United States of America

About the Book

Olduvai, Laetolil, Omo, Turkana, and Hadar are now nearly household words in the lexicon of early man. Eastern Africa, from Transvaal through Tanzania, Kenya, and Ethiopia, has held a near monopoly on hominid discoveries. But in recent years, archaeological geologists have joined the anthropologists and paleontologists in efforts to broaden the search for early man--defining the environmental parameters of both a habitat capable of sustaining human development and the geological conditions conducive to the preservation of faunal and artifactual remains, thus helping to ascertain more productive places to search for early hominid sites.

The authors of this book provide detailed accounts of several of these efforts in archaeological geology. Showing how geochronology, sedimentology, petrology, pedology, palynology, geophysics, zooarchaeology, and related studies have expanded our knowledge of hominid development, they cover work at many of the better-known sites, as well as at a potential hominid site in the Philippines.

About the Series

The *AAAS Selected Symposia Series* was begun in 1977 to provide a means for more permanently recording and more widely disseminating some of the valuable material which is discussed at the AAAS Annual National Meetings. The volumes in this *Series* are based on symposia held at the Meetings which address topics of current and continuing significance, both within and among the sciences, and in the areas in which science and technology impact on public policy. The *Series* format is designed to provide for rapid dissemination of information, so the papers are not typeset but are reproduced directly from the camera-copy submitted by the authors. The papers are organized and edited by the symposium arrangers who then become the editors of the various volumes. Most papers published in this *Series* are original contributions which have not been previously published, although in some cases additional papers from other sources have been added by an editor to provide a more comprehensive view of a particular topic. Symposia may be reports of new research or reviews of established work, particularly work of an interdisciplinary nature, since the AAAS Annual Meetings typically embrace the full range of the sciences and their societal implications.

WILLIAM D. CAREY
Executive Officer
American Association for
the Advancement of Science

Contents

About the Editors and Authors

George Rapp, Jr., *is dean of the College of Letters and Science, director of the Archaeometry Laboratory, and professor of geology and archaeology at the University of Minnesota-Duluth. Among his numerous publications are* Troy: The Archaeological Geology *(J.A. Gifford, coeditor; Princeton University Press, in press),* Excavations at Nichoria in Southwestern Greece: Site, Environs and Techniques, Vol. 1 *(S.E. Aschenbrenner, coeditor; University of Minnesota Press, 1978), and* The Minnesota Messenia Expedition: Reconstructing a Bronze Age Regional Environment *(W.A. McDonald, coeditor; University of Minnesota Press, 1972).*

Carl F. Vondra *is a professor of geology at Iowa State University. He is currently involved in studies of the Pleistocene geology of the Cagayan Basin (Northern Luzon, Philippines) and the East Turkana Basin (Northern Kenya) and the drainage history of the Rhine River during the Late Cenozoic. He has also conducted stratigraphic research in the Fayum Depression in Egypt, the Siwalik range in northern India, and in selected basins in the western United States.*

R. H. Ardrey *is a senior geologist with Husky Oil International, Ltd. His interests are in sedimentation and stratigraphy, and he has published on the diagenesis of the Middle Ordovician Trenton formation in southern Michigan.*

James L. Aronson, *associate professor of geology at Case Western Reserve University, has specialized in geochronology and geology in areas of young volcanic activity and mountain-building. He has studied the "island continents" of the Southwest Pacific, volcanic activity in Iceland, and the geochronology and geology of the Hadar hominid site in Ethiopia.*

Russell B. Bainbridge, Jr., *is Oil and Mineral Property Manager for the Continental Bank in Chicago. A specialist in sedimentation and stratigraphy, he has done field research in East Turkana, Kenya.*

Francis H. Brown, *professor of geology and geophysics at the University of Utah, is a specialist in tetrology and geochronology. He has published on the geological background of fossil man, the calibration of hominoid evolution, and the chronology and correlation of Plio-Pleistocene East African hominid sites.*

Daniel R. Burggraf, Jr., *a research associate completing his doctoral dissertation at Iowa State University, has done research on sedimentation, stratigraphy, and volcaniclastics. He has published on fluvial sedimentology and fluvial subfacies at Lake Turkana, Kenya, and is currently studying volcaniclastic sediments in the Philippines, East Africa, and the western United States.*

Hal J. Frank, *a project geologist for the Continental Oil Company in San Antonio, has done field research in East Africa.*

Richard L. Hay, *a specialist in stratigraphy and sedimentary petrology, is a professor of geology at the University of California-Berkeley. His emphasis has been on the mineralogy and chemical reactions in soils and lake sediments, and he is the author of* Geology of the Olduvai Gorge *(University of California Press, 1976). His work has earned him the Arnold Guyot Award of the National Geographic Society and the Kirk Bryan Award of the Geological Society of America.*

Gary D. Johnson *is an associate professor of earth sciences at Dartmouth College. He has published on fossil soils, paleomagnetic stratigraphy, molasse sedimentation, and fission-track geochronology. He has conducted fieldwork at hominid sites in Kenya, India, and Pakistan.*

Erik P. Kvale, *a graduate assistant in geology at Iowa State University, is currently studying the stratigraphy of the Plio-Pleistocene sediments in the Cagayan Valley, Northern Luzon, Philippines. His previous studies include fieldwork in Europe and the Canary Islands.*

Mark E. Mathisen *is a doctoral candidate in geology at Iowa State University and an honorary researcher for the Philippine National Museum. His area of specialization is stratigraphy and sedimentation, and he has written on the provenance and environment of the Plio-Pleistocene sediments in the East Turkana Basin, Kenya.*

Neil D. Opdyke, *a senior research associate at Lamont Doherty Geological Observatory, is well known for his work in paleomagnetism. He has served on the IUGS Subcommittee on the Magnetic Polarity Time Scale, and he is currently studying the paleomagnetic stratigraphy of the Neogene of Pakistan.*

Pamela H. Rey *is a geological engineer at Woodward-Clyde Consultants in San Francisco.*

R. A. Khan Tahirkheli *is professor and director of the National Centre of Excellence in Geology, University of Peshawar, Pakistan. He has three decades of practical experience as a geologist, and among his publications are two books which he recently edited:* Geology of Kohistan, Karakoram Himalaya, Northern Pakistan *and* Proceedings of the International Committee on Geodynamics, Group 6 Meeting at Peshawar *(23-29 November 1979).*

Maurice Taieb *is Chargé de Recherche and Chargé de Cours in the Laboratoire de Géologie, Centre National Recherche Scientifique, France; his areas of specialty are stratigraphy, cartography, sedimentology, and paleomagnetism. He organized and was director and chief geologist of the International Afar Research Expedition and is a research associate at the Cleveland Museum of Natural History.*

Charles F. Visser *is a project geologist for Amoco Production Company. His fields of interest include molasse sedimentation and the structural geology of thrust-fold belts. His fieldwork includes studies of the sedimentology, stratigraphy, and tectonic control of Siwalik molasse in Pakistan.*

Howard J. White *is completing his doctoral dissertation in geology at Iowa State University. His research includes paleoenvironmental and basin analysis of the western United States (Cenozoic), a paleosol analysis of the Plio-Pleistocene sediments in the Eastern Turkana Basin (Kenya), an examination of the Plio-Pleistocene boundary problem, and tectonics and sedimentation of the Indo-Pakistan collision zone.*

Acknowledgments

The AAAS symposium and this volume originated through an invitation by Ray Bisque, secretary of Section E, Geology and Geography. Coeditor Carl Vondra, long one of the key archaeological geologists in hominid studies, has provided the necessary guidance and geological judgment in bringing this volume together.

Rebecca Judge, assisted by Sheila Arimond, undertook the arduous task of careful editing of the manuscript copy and illustrations. Karin Cowper, Sherri Fedora, Marian Syrjamaki, and Vi Wangensteen typed the camera-ready copy, and Vicki Spragg and Donald Batkins of the Cartography Lab at the University of Minnesota, Duluth, prepared the final versions of many of the illustrations. Joellen Fritsche has been responsible for AAAS editorial management.

To all of these I want to express my thanks and appreciation for making this volume possible. I retain for myself the responsibility for errors, omissions, and oversights.

George Rapp, Jr.
Archaeometry Lab
Duluth, Minnesota
August 1980

Introduction

Africa, from Transvaal through Tanzania, Kenya, and Ethiopia, has held a near-monopoly on hominid discoveries. Here and in southeastern Asia, archaeological geologists have joined with paleoanthropologists and paleontologists to broaden our understanding by describing the sequence of natural habitats that both sustained (and probably helped govern) evolutionary change and provided geological conditions conducive to the preservation of biological and artifactual remains. Their efforts have expanded the geography and geology of hominid research as the papers in this volume attest.

Relevant factors in the paleoenvironment that markedly affected hominid development can be grouped under three major headings:

Geology: including physical geography, pedology, distribution and frequency of volcanism and tectonic movements, the hydrologic regime, the strength and polarity of the earth's magnetic field, and nature of available lithic materials.

Climate: primarily the amount and seasonal distribution of precipitation and temperature.

Biology: the fauna and flora in equilibrium in a regional ecological setting will depend primarily on the geology and the climate.

The sequence of geologic environments and events is recorded in the sedimentary deposits that form at or near the earth's surface. Such deposits often contain the biological and climatic as well as the geologic record. The time scale involved in hominid evolution allows for extensive physiographic and climatic changes. These constrained the possibilities of hominid development and the growth of hominid cultural complexity.

Geologic strata provide a chronological as well as an ecological record. The tuffs of the East African Rift System provide numerous marker beds amenable to potassium/argon or fission track dating.

It should be stressed that the period of development of the advanced hominids was a time of repeated ecological shifts such as the expansion and contraction of continental ice sheets which affected sea level and climate far beyond the borders of the ice masses themselves. At least fifteen major Pleistocene climatic fluctuations can be recognized in deep sea cores. The development of the genus Homo took place during this unique period in earth history. Although the glaciers did not reach most of southern Europe, southeast Asia or Africa during glacial expansion, the vegetational zones were compressed toward the equator and temperature patterns fluctuated with the waxing and waning of the large ice masses.

Somewhat before the onset of the Pleistocene, a group of Lower Paleolithic hominids began to manufacture tools. Culturally inheritable tool-making skills distinguished these primates from their contemporaries and provided an advantage in the efficiency of food procurement and opportunities for working materials, an important step in evolution. This transmission of tool-making skills from generation to generation may have played a role in the development of speech.

While the archaeological record is still too inadequate for definite statements on the early evolution of the genus Homo, in the last two decades sufficient hominid sites have been discovered to begin detailed paleoecological analyses of the environmental settings of hominid habitats. These analyses are drawn from the disciplines and subdisciplines of geology (geophysics, sedimentology, geomorphology, paleontology, volcanology, geochronology, archaeological geology), paleozoology, paleobotany, osteology, and related fields.

If we are to speak extensively and specifically about environmental influences on the course of human evolution we need to know more about the details of hominid evolution. Archaeological geology can provide a necessary supplement to this endeavor by describing the environmental circumstances surrounding hominid evolution. Field work, in exceptional cases, even turns up records such as the famous Laetoli hominid footprints.

Terrestrial strata record evidence of the surface water regimes which in conjunction with the many wet/dry indicators from the floral and faunal remains can be used to reconstruct ecological gradients through time. Rapid expansion or contraction of humid, forested, lacustrine environments drastically affect the ecological carrying capacity of a region. When associated with major geologic change such as tectonic movements and volcanism, purturbation of the established environmental equilibrium could well instigate extinctions or discontinuous evolutionary adaptations.

We know that hominids ranged widely utilizing a variety of landscape habitats. As congenial habitats were destroyed by geologic factors, ecologic stress invariably descended on the entire biological community. Whatever their influence, profound environmental modifications appear to have accompanied hominid evolution.

The geography of hominid distribution was affected by a variety of geologic events. Eustatic sea-level changes accompanied by isostatic depression and recovery controlled land bridges between islands and continents. Periodic Pleistocene glaciations withdrew, then later returned over 100 m of water uncovering or flooding large epicontinental areas such as northwestern Europe and southeastern Asia. These sea level changes were accompanied by shifting floral and faunal zones. Each major fluctuation necessitated some cultural response, such as migration, for affected hominid groups.

In tropical and equatorial regions often one significant result of glaciation was increased rainfall, resulting in the expanded development of river systems. It was during the Middle Pleistocene that hand axes appeared. Hand axe makers, adept at big game hunting, concentrated in the valleys of the large river systems in Europe, Africa and Asia. Olduvai Bed IV has provided evidence of a major hand axe industry.

As taphonomic studies have shown, different geologic environments vary not only in their capacity to preserve remains of past ecological systems but also in how they affect the nature and extent of what is preserved. Diagenesis, pedogenesis and bioturbation serve to promote differences between fossil assemblages (including pollen and phytolith assemblages) and the living communities from which they were derived. Hence any effort to correct the resulting biases requires a thorough knowledge of the post-depositional history of the deposits.

Another limiting factor in the geologic record is its incompleteness. Few areas of the earth's land surface have witnessed long and continuous periods of deposition. Indeed erosion is the terrestrial norm. Lakes are ephemeral features of the landscape and rivers meander back and forth in broad valleys destroying much of their earlier record. Hence time/stratigraphic gaps in the depositional record prevent detailed paleoecological reconstructions for important segments of time. Unfortunately the early terrestrial record of the early Pleistocene is poor and often disturbed by later Pleistocene events. Only in East Africa do we have much detailed Lower Pleistocene stratigraphy. Geologic field work in progress in southeast Asia should add measurably to the record.

This AAAS symposium volume presents seven in-depth accounts of the geologic parameters affecting the paleoecology of important hominid sites or sites of potential significance for hominid discoveries. Chapters were written by specialists who have been responsible for significant work in the region described. The authors of this volume have built on the work of H. Reck, C. Albritton, B.H. Baker, K. Behrensmeyer, W.W. Bishop, A. Brock, K. Butzer, J.D. Clark, Y. Coppens, F.C. Howell, G.L. Isaac, D.C. Johanson, M.D. Leakey, L.S.B. Leakey, R.E. Leakey, D. Pilbeam, F. White, L.A.J. Williams, and others.

The individual chapters vary in their scope, depth and technical difficulty. Yet the wealth of information contained in this symposium volume contributes significantly to the rapidly expanding literature on the paleoenvironmental settings of hominid sites.

We begin in East Africa where a mid-Miocene forest and woodland landscape has been modified by uplift, rifting, and volcanism accompanied by a very long-term decrease in average annual temperature, to include many additional environmental niches from semiarid desert to tropical forest. Only in Africa have hominid fossils been recovered for the important period dating between five and two million years ago.

For nearly two decades Richard L. Hay has developed the geologic framework for an understanding of the hominid discoveries of the Leakeys' in East Africa. In his opening chapter on the Laetolil Beds he presents the geologic evidence for a semiarid and grassland savanna environment similar to that of the Laetoli area today. Unlike most of the other hominid-containing deposits which are lake or river margin deposits, the Laetolil Beds are dominated by windblown sand and ash. Hay's careful work has delineated the volcanic, tectonic and geomorphic features governing the ancient landscape. The Laetolil Beds contain a rich vertebrate fauna, including hominids, and volcanics that yield to potassium/argon dating of the depositional sequence.

Hay's work on Bed I of Olduvai Gorge on the edge of the Serengeti Plain, published in Quaternary Research in 1973, still stands and we have reprinted it with permission in this volume. He defines five depositional environments and produces evidence for the related paleogeography, chronology, biota, and climate, thus providing a detailed picture of the habitats encompassing the hominid activity.

This volume turns next to the rift valley settings in East Turkana with two fine chapters by Carl F. Vondra and his associates. The whole period of hominid development in East Africa has been marked by the dynamic tectonic regime of the East African Rift System and the accompanying volcanism. The effect of the faulting on the depositional environments, the related landscapes and the favorable geologic conditions for the preservation of hominid remains are clearly presented by the authors.

Still further north in the rift corridor that runs from Olduvai through Lake Turkana is the lower valley and basin of the Omo River which contains a thousand meters of sediments deposited during the period of hominid evolution. Unlike the regions far to the south where tectonic forces developed major regional uplift, there is no geologic evidence in the Omo basin for large altitude changes in the last 3.5 m.y. Paleontological work on the Omo deposits has described a rich vertebrate fauna. Francis H. Brown, in his chapter on this region, has delineated the climatic and ecological regimes during the deposition in the basin of the extensive Shungura Formation.

The oldest "man-made" artifacts come from transitional lake sediments, dating to 2.5 m.y. B.P. at Hadar on the western margin of the Afar region of Ethiopia. As with many hominid finds in East Africa, those from the Hadar Hominid Site can be dated by a sequence of volcanic tuffs. Authors James Aronson and Maurice Taieb have examined the geology and paleogeography of this region, and have described in their chapter the main geologic parameters governing the paleoecology of this region during the Pleo-Pleistocene transition.

Although the deposits in East Africa have been the focus of hominid exploration in the last three decades, the 7,000 meter-thick Siwalik Group exposed to the southwest of the Himalayas has great potential for providing major evidence

for our understanding of hominid development in the critical
period from eight to six million years ago. Deposits from
the period of eight to five million years ago have been
essentially devoid of hominid fossils. And this is a crucial
period in the background to the evolution of the genus <u>Homo</u>.
Gary D. Johnson and his colleagues have written a highly
technical and detailed account of the geology and pedology of
this important rock sequence. The relation of climate to
soil development is well illustrated in this chapter.

In the last chapter of this volume we concentrate on the
Philippines, another region of southeast Asia with potential
for important hominid discoveries. Here again Carl Vondra
and his group report on the results of a carefully designed
and executed program of geologic, archaeological, and paleontological exploration in the Cagayan Valley, Northern Luzon.
Scattered stone tools associated with strata yielding a
varied mammalian fauna give promise of hominid remains.
Discoveries in this region would have the added advantage of
volcanism concommitant with the deposition of deltaic and
fluvial sediments. As in East Africa these volcanic deposits
offer the possibility of closely dating many of the strata.

1. Paleoenvironment of the Laetolil Beds, Northern Tanzania

Abstract

The Laetolil Beds of northern Tanzania contain remains and footprints of Pliocene hominids. These deposits are as much as 130 m thick and are subdivided into two units, the upper of which is 45 to 60 m thick and contains a rich faunal assemblage including the hominids. These beds have been dated by K-Ar at 3.5 to 3.7 m.y. B.P.

The upper, fossiliferous part of the Laetolil Beds consists almost entirely of land-laid tuffs, most of which were extensively reworked by wind. These tuffs were deposited on a land surface of low relief and were weathered before burial to constitute a sequence of weakly developed paleosols of caliche type.

The faunal assemblage is similar in most respects to that of the present time in the same area and clearly points to a grassland savanna lacking major rivers or nearby lakes. Both fauna and lithology indicate that the climate was semiarid, and the footprint tuff suggests well-defined dry and rainy seasons as at present. On the basis of Heterocephalus (naked mole rat) in the Laetolil Beds, the temperature may well have been warmer than at present.

The Laetolil Beds were deposited on the crest and flanks of a broad domal uplift formed prior to 3.7 m.y. B.P. Domal warping was repeated about 2.5 m.y. B.P., shortly prior to a major episode of volcanism 2.1 to 2.4 m.y. B.P. Rift-valley faulting began about 2.1 m.y. B.P., forming the Eyasi and Olduvai basins and the Eyasi Plateau.

Introduction

The Laetolil Beds presently represent the oldest well-documented hominid site in East Africa. The upper, fossil-

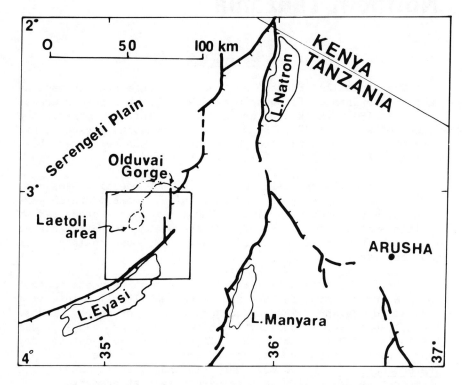

Figure 1. Regional map showing the location of the
Laetoli   area and   Olduvai Gorge   within
the   northern   Tanzanian   sector of   the
eastern rift zone.

iferous part of the Laetolil Beds has been dated at 3.5 to
3.m.y. by the K-Ar method using biotite from tuffs (Leakey,
and others, 1976). The first hominid fossil, a maxilla, was
found in 1938-39 by the expedition of Kohl-Larsen (Kohl-
Larsen, 1943), and numerous remains, principally mandibles
and teeth, have been found by the expeditions of M.D. Leakey
from 1975 to 1979 (Leakey, and others, 1976; Johanson and
White, 1979).   Of special interest are the trails of hominid
footprints found in a tuff bed (Leakey and Hay, 1979). This
site is ecologically significant as the only upland East
African hominid site presently known.   Other East African
hominid sites are chiefly in lake-margin and riverine settings.

   The Laetolil Beds are located on the southern part of the
eastern rift valley, where the central graben has splayed out
into a series of tilted fault blocks with lake basins on
their downthrown sides and plateaus on their upthrown sides
(Fig. 1).   The Laetolil Beds are exposed over an area of at
least 1500 km^2 on the upthrown fault block to the north of
Lake Eyasi. These deposits overlie Precambrian basement
rocks and are bordered and overlain on the east by volcanic
rocks of several large volcanoes (Fig. 2). The Laetolil Beds
were first described and named by Kent (1941), and
Pickering's mapping (1964) extended their known distribution.

   The richly fossiliferous exposures of the Laetolil Beds
are confined to an area of about 70 km^2 on and near the
drainage divide between the Eyasi lake basin and Olduvai
Gorge (Fig. 2).   This area was earlier designated Laetolil
(Kent, 1941) but has been renamed Laetoli, which is the more
correct anglicization of the Masai word of this area.   The
Pliocene deposits will continue to be known as the Laetolil
Beds.

   The Laetolil Beds in the Laetoli area generally crop out
in a series of shallow, discontinuous exposures, but a section 130 m thick was measured in one place. The base of the
section is not exposed, and the Laetolil Beds may well be
substantially thicker than this.   The Laetolil Beds are
disconformably overlain by the Ndolanya Beds, which are 14 to
20 m thick, fossiliferous, and consist principally of eolian,
or wind-worked tuffs. The Ndolanya Beds have thus far proved
unsuitable for radiometric dating but are bracketed in age
between the Laetolil Beds and lavas dated at 2.4 m.y. The
lavas are overlain by relatively unfossiliferous trachytic
tuffs named the Naibardad Beds and correlated on the basis of
mineralogy with at least the lower part of Bed I in Olduvai
Gorge, which is 1.7 to about 2.1 m.y. old (Hay, 1976). The
next younger deposits are the Olpiro Beds, which are repre-

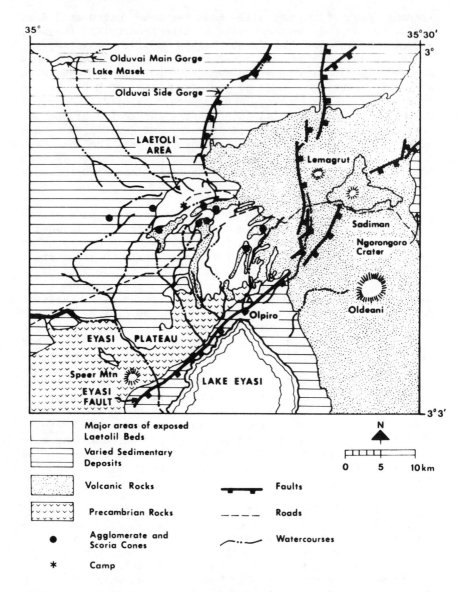

Figure 2. Geologic map showing the main areas of exposed
Laetolil Beds and the major associated geologic
units and physiographic features. Map unit
"Sedimentary deposits of varied ages" consists
chiefly of late Pleistocene deposits but includes
minor exposures of the Laetolil Beds. Map unit
"Major areas of exposed Laetolil Beds" also
includes younger deposits in the Laetoli area.

sented by discontinuous, widely scattered exposures, chiefly of claystone. These beds contain faunal remains and arti-facts similar to those of middle Bed II of Olduvai Gorge (M.D. Leakey, 1976, personal communication), which has an age of about 1.3 to 1.4 m.y. The Ngaloba Beds, described by Kent (1941), are the youngest stratigraphic unit in the Laetoli area. Tuffs of the Ngaloba Beds are similar to those of the Ndutu Beds of Olduvai Gorge (~40,000 to 400,000 yr. B.P.). In the Laetoli area prior to 1975, fossils from all deposits older than the Ngaloba Beds were assigned to the Laetolil Beds, causing considerable paleontologic confusion (Cooke and Maglio, 1972; Leakey and others, 1976).

Stratigraphy, Lithology, and Origin of the Laetolil Beds

Stratigraphy and Correlation

The Laetolil Beds in the Laetoli area are subdivided into two units, the lower of which is at least 70 m thick and con-sists of water-worked and eolian tuffs and mudflow deposits. The upper unit, 45 to 60 m thick, consists principally of eolian and air-fall tuffs. The following discussion will be confined to the upper unit, which has yielded nearly all of the vertebrate fossils.

The upper unit of the Laetolil Beds is about 80 percent reworked and 20 percent air-fall tuff (Fig. 3). Nearly all of the reworked tuffs are of air-fall ash redeposited by wind and are termed eolian tuff. Perhaps 1 to 2 percent of the tuffs are stream-worked. The air-fall tuffs comprise a thick tuff of nepheline phonolite composition and numerous thin tuffs of melilitite-carbonatite composition (Hay, 1978). The more distinctive and widespread of the tuffs, numbered 1 to 8, were used for correlating. Horizons of lapilli and blocks of lava and plutonic rock assist in correlations.

Description and Origin

Eolian tuffs are massive to crudely bedded, typically well-cemented, and gray or brown and similar in most respects to the eolian tuffs of Olduvai Gorge (Hay, 1976). They are poorly sorted to moderately well-sorted and dominantly of medium-sand size. Ash particles consist of volcanic glass, rock fragments, and crystals representing eruptions of low-silica magma of nephelinite and melilitite composition. Half or more of these particles have pelletoid coatings of very fine-grained tuff or clay. The coatings were formed by accretion prior to burial and characterize some ash deposits that have been extensively redeposited by wind (Hay, 1976;

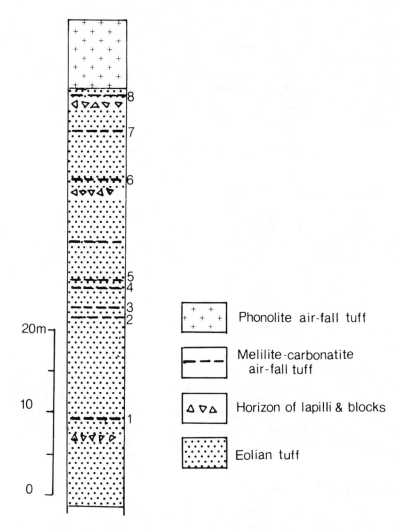

Figure 3. Columnar section of the upper, fossiliferous unit of the Laetolil Beds in the northeastern part of the Laetoli area.

Hay and Reeder, 1978). These eolian tuffs are cemented by calcite, phillipsite, montmorillonite, and less commonly by dolomite (Hay, 1978).

Fossil termitaries and evidence of bioturbation are widespread and locally abundant throughout the upper unit of the Laetolil Beds. Termitaries are of several types, the most common of which is an irregular network of branching cylindrical channels filled by calcite-cemented tuff (Fig. 4). Water-worked tuffs form lenticular deposits filling stream channels, which are on the order of 30 to 100 cm deep and 1 to 3 m wide.

The melilitite-carbonatite air-fall tuffs are unusual in several respects and are described below. The thick nepheline phonolite air-fall tuff will not be considered further. The melilitite-carbonatite tuffs consist largely of crystals and globule-shaped lava particles 0.2 to 1.0 mm in diameter cemented by calcite. These tuffs originally contained a substantial proportion of carbonatite ash, which is an extraordinary material consisting chiefly of anhydrous sodium and calcium carbonate. Most tuffs are between 1 and 15 cm thick, unstratified, and represent single ash falls; a few are 15 to 100 cm thick, laminated, and represent a series of closely spaced ash showers.

A polygonal fracture pattern (Fig. 5) characterizes nearly all of the tuffs less than 10 cm thick. The tuffs were fractured and the polygonal tuff clasts, rounded and locally displaced before burial by eolian tuffs. Displaced and rounded tuff polygons show that the tuffs were cemented soon after deposition, while the underlying eolian tuffs were still unconsolidated. Polygonal fracturing resulted largely from movements, particularly bioturbation and expansion and contraction related to wetting and drying, in uncemented eolian sediment underlying the cemented air-fall tuffs. Tuffs were very likely cemented initially by the sodium salt trona ($Na_2CO_3 \cdot NaHCO_3 \cdot 2H_2O$) which crystallized from a solution of sodium carbonate dissolved from carbonatite ash by rainfall or dew (Hay, 1978). This "instant cementation" accounts, at least in part, for the excellent preservation of footprints at many horizons in the Footprint Tuff, which represents a series of thin, closely spaced ash falls (Leakey and Hay, 1979). The tuffs were later cemented to form a hard rock by calcite and phillipsite.

Air-fall tuffs are generally flat or gently undulating, and the maximum local relief is about 2 m over a distance of 20 m. Tuffs are continuous, even where they overlie hillocks, thus providing another line of evidence for early

Figure 4. Photograph showing the irregular network of channels which represents the dominant type of termitary structure in eolian tuffs of the Laetolil area.

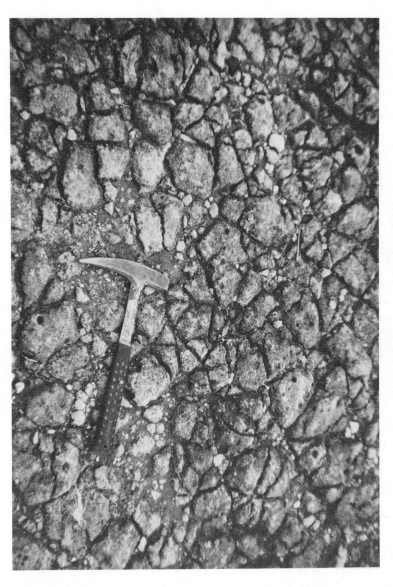

Figure 5. Photograph of air-fall tuff no. 6 (Fig. 3) showing the polygonal fracture pattern characteristic of the thinner air-fall tuffs in the Laetolil Beds.

cementation. Wind or sheetwash of heavy showers would have eroded the ash from hillocks if it had not been cemented soon after it was deposited.

Paleo-environment Inferred from Lithology

Eolian tuffs indicate an environment in which vegetation was at least seasonally insufficient to prevent extensive transportation of sand-sized ash particles by wind. The eastern semiarid-to-arid part of the Serengeti Plain provides a modern example of comparable wind-worked ash (Hay, 1976). This area is a grassland savanna with scattered brush and Acacia trees. Rainfall is highly seasonal and averages 50 cm per year. The plain is blanketed with a deposit of ash, most of which was erupted from the active volcano Oldoinyo Lengai about 1300 yr. B.P. This ash deposit has been extensively redeposited by wind, and it is still actively reworked during the dry season. Dunes of reworked ash move during both the wet and dry seasons.

Tuffs of the Laetolil Beds were weathered to varying extents before burial and can be considered a sequence of weakly developed paleosols. Minerals formed by weathering are calcite, montmorillonite, phillipsite, and dolomite. Calcite is the most abundant, showing that the paleosols are of caliche type. The nature of the weathering products is evidence of a saline, alkaline soil environment such as is found today in the eastern part of the Serengeti Plain. The soil environment very likely had a pH of 9.5, at least seasonally, as a result of a dry climate and carbonatite ash, as on the Serengeti Plain (Hay, 1978).

Paleontology

Faunal Remains

The Laetolil Beds contain a wide variety of fossil vertebrates (Kent, 1941; Dietrich, 1942; Leakey and others, 1976). Remains collected in 1975 are listed in order of decreasing frequency as follows: bovids (especially Madoqua or dik dik), lagomorphs (rabbits), giraffids, rhinoceros, equids, suids, proboscideans, rodents, carnivores, and primates (Leakey and others, 1976). Reptiles are represented by snake vertebrae and tortoises, and avian remains, including bird eggs, occur widely. Land snails are both common and widespread, as are mantles of slugs (cf. Urocyclids).

This assemblage is characteristic of an upland savanna, which is further supported by the absence of hippopotamus, crocodile, and other water-dwelling forms. The fauna is in

most respects similar to that of the same area at the present
time.   The small rodents are especially important for a
paleoenvironmental reconstruction because they do not migrate
through the year, unlike the larger animals, and some of them
occupy restricted ecological niches.   The rodent fauna has
been characterized by J. J. Jaeger as follows (1977, personal
communication):

> "Rodents collected from several levels through
> the Laetolil formation are represented mainly by
> remains of spring-hare (Pedetes), pouched-mouse
> (Saccostomus), naked mole rat (Heterocephalus) and
> ground squirrels (Xerus). This association, with
> the abundance of the hare (Serengetilagus), strongly
> suggests an open grassland environment. The existence
> of thorn trees is also indicated by the occurrence
> of the tree-rat Thallomys."

Heterocephalus, the naked mole rat, provides tentative
evidence that the climate was warmer than in the Laetoli
area today.   The Laetoli area lies at an elevation of about
1800 m, and the temperature commonly drops to 5° to 6°C at
night.   The modern species of Heterocephalus cannot tolerate
temperature this low (J. J. Jaeger, 1976, personal communication),   and   if Pliocene Heterocephalus had   the same
tolerance, then night-time temperatures must have been significantly warmer.   Warmer temperature could have resulted
from a lower elevation, which is independently suggested by
the structural history of the area outlined below.

Footprints

Fossil footprints have been found in tuffs 3, 7, and 8
(Fig. 3) but are most abundant in the lower 15 cm of Tuff 7,
which is termed the Footprint Tuff.   Footprints identified
thus far in the Footprint Tuff can correspond rather closely
with forms represented by faunal remains.   Among the most
common footprints are those of lagomorphs, rhinoceros,
bovids, proboscideans, and birds (cf. guinea fowl).   Of special interest are those of Papio (baboon) and hominids.

The Footprint Tuff represents a series of thin ash falls
which accumulated over a period of a few weeks to perhaps as
much as several months.   The lower 7 to 9 cm of the Footprint
Tuff is composed of laminae 1 to 5 mm thick which exhibit
mantle bedding and extend widely over the Laetoli area. Rain-
prints made by light showers are found at several horizons
and aid in correlating individual tuff laminae.   The overlying tuffs lack mantle bedding and exhibit small-scale
cross-bedding and other features showing that they had been

Figure 6. Air view of the Eyasi fault scarp with Lake Eyasi in upper
right. View was taken looking northeast from the vicinity
of Speer Mountain, shown near the south edge of Figure 2.
Fault scarp in foreground is about 800 m high.

extensively redeposited by water. The lack of any erosional hiatus within the Footprint Tuff suggests that it
accumulated over a relatively short period, certainly less
than a year. Using the present-day climate as a model,
the lower part of the Footprint Tuff is believed to have
been deposited during the latter part of the dry season,
when light showers are not infrequent, and the upper unit
was deposited during the rainy season (Leakey and Hay, 1979).

The stratigraphic pattern of footprint types supports the
dry- to rainy-season hypothesis. Footprints in the lower
part are chiefly rabbit, rhinoceros, guinea fowl, bovids, and
other forms which remain in grassland savanna during the dry
season. Footprints in the upper part include a wide variety
of forms not found in the lower part (e.g., proboscideans,
equids, Papio, and hominids). These forms found only in the
upper part of the tuff may represent the rainy-season migration, which is a characteristic feature of the East African
savannah.

Paleo-environmental Summary

1. The Laetolil Beds in the Laetoli area were deposited
on an upland surface of low relief.
2. The eolian tuffs required an environment with vegetation either sparse or low enough to permit extensive transport of ash particles by wind.
3. The faunal assemblage indicates a grassland savanna
with Acacia trees. The lack of hippopotamus and crocodile
remains supports other evidence for an upland terrain lacking
major rivers or nearby lakes.
4. The climate was semiarid as shown by the faunal evidence and the mineralogic nature of paleosols. The climate
was probably warmer than that in the same area today.
5. The climate very likely had well-defined dry and
rainy seasons, as at the present time.

Regional Paleogeography and
Structural History

Structural Setting

The bulk of the Laetolil Beds are located on the Eyasi
Plateau, which is on the large block uplifted along the Eyasi
fault (Fig. 6). The Laetoli area is at the southern margin
on the plateau, at an elevation of about 1800 m on the
watershed between the Olduvai and Eyasi drainage basins. The
Eyasi fault scarp increases in elevation from about 300 to
800 m from northeast to southwest.

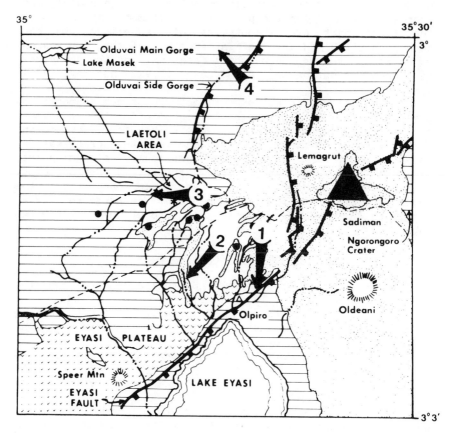

Figure 7.   Direction of stream flow recorded in the Laetolil
Beds and   superimposed on the geologic map of
Figure 2.   Solid triangle represents the volcano
Sadiman,   the   probable   source   of the Laetolil
Beds.   Arrow 1 represents a 300-m-deep valley
filled   by   the   Laetolil   Beds.   Arrows 2 and 3
represent the drainage direction determined from
stream-channel   orientation in the Laetolil Beds,
and arrow 4 is inferred from the location and
lithology of a streamlaid facies of the Laetolil
Beds.

Kent (1941) and Pickering (1964) inferred that the Eyasi fault scarp and plateau existed when the Laetolil Beds were deposited. Kent seems to have drawn his conclusion from the upland nature of the Laetolil fauna, whereas Pickering based his conclusion on a steep-sided valley exposed in the fault scarp and filled by the Laetolil Beds. The valley is about 300 m deep and is located to the north of Olpiro (Fig. 2), near the northeast end of the fault scarp. He interpreted the filled valley as cut through the scarp with its present relief.

Evidence presented here shows, to the contrary, that development of the scarp commenced about 2.1 m.y. B.P. The filled valley noted by Pickering is here interpreted as evidence that the scarp did not exist when the valley was filled by a 300 m thickness of Laetolil Beds. The Laetolil Beds filling the valley are chiefly waterlaid tuffs and mudflow deposits (M. Monaghan, 1976, personal communication). If the valley had been cut into the scarp as at present, then the Laetolil Beds would not have filled the valley but rather would have spread out as a fan at the mouth of the valley, filling no more than its lower part. As inferred here, the valley was cut into an upland surface extending a considerable distance to the south of the present-day scarp. It was filled by the Laetolil Beds, which were later partly eroded from the valley. Channels cut into the Laetolil Beds were filled about 2.1 to 2.4 m.y. B.P. by lava of the volcano Lemagrut. These lavas slope at a gentle angle to the south and are abruptly terminated by the scarp, thus demonstrating that faulting followed eruption of the lavas. The age of Lemagrut is estimated at 2.1 to 2.4 m.y. on the basis of its stratigraphic relations to dated rocks of Ngorongoro (Hay, 1976) and the fact that Lemagrut flows overlie lava from vents in the Laetoli area that has been dated at 2.4 m.y. (Leakey and others, 1976).

The drainage pattern recorded in the Laetolil Beds (Fig. 7) shows that they were deposited over a broad domal uplift prior to movements on the Eyasi fault. The 300-m-deep valley filled by the Laetolil Beds indicates drainage from north to south (Fig. 7, arrow 1). Seven kilometers west of the filled valley, streams flowed from northeast to southwest as determined from the orientation of 15 stream channels within a fluvial facies of the lower unit of the Laetolil Beds (Fig. 7, arrow 2). This flow direction, parallel to the fault scarp, is perpendicular to the present-day streams in the same area, which flow directly toward the fault scarp. These Laetolil stream channels argue strongly against the existence of the Eyasi fault scarp. Streams in the Laetoli area flowed to the west, as determined from the orientation of 14 stream channels in the upper unit of the Laetolil Beds. To the north, in the Olduvai Side Gorge, a stream flowing roughly northwest is inferred on the basis of a fluvial facies of the upper unit which contains blocks of lava as much as 30 cm in diameter. The large size of the blocks is compatible with a relatively high-velocity, low-sinuosity stream originating on the slopes of Sadiman, the source volcano. These data, though few, seem to represent radial drainage on the western part of a domal uplift.

Small-scale localized doming, or anticlinal folding, took place following deposition of the Ndolanya Beds along an east-west axis passing through the northern part of the Laetoli area. The upper unit of the Ndolanya Beds contains **Hipparion** cf. ethiopicum, whose age is estimated at about 2.5 m.y. (Hooijer, 1979). Lavas dated at 2.4 m.y. were erupted after the episode of folding, which can thus be dated at about 2.4 to 2.5 m.y.

The modern drainage pattern, with streams flowing south or southwest toward the fault scarp, was developed in the Laetoli area shortly prior to the deposition of the Naibardad Beds. This drainage pattern was developed, at least to some extent, in response to the Eyasi fault scarp. Thus the first significant movements on the Eyasi fault and development of the fault scarp can be dated at about 2.1 to 2.2 m.y. on the basis that Lemagrut lavas ($\sim$2.1 to 2.4 m.y.) exposed in the fault scarp were erupted prior to faulting, and the oldest deposits of the Naibardad Beds ($\sim$2.1 m.y.) were deposited after faulting had begun. The Naibardad Beds consist of tuffs erupted from Ngorongoro, thus demonstrating that rift-valley faulting commenced during the major phase of volcanism which produced the volcanoes Lemagrut, Ngorongoro, and Oldeani.

The Olduvai Basin was formed about 2.1 m.y. B.P., at about the same time as the Eyasi scarp and lake basin. It seems quite likely that the Olduvai and Eyasi basins were produced by the same phase of faulting in view of their proximity and interconnecting fault systems (Figs. 1 and 2). Movements continued intermittently along faults in the Olduvai Basin (Hay, 1976) and possibly along the Eyasi fault until very late Pleistocene time.

Doming followed by rift-valley faulting has been docu-mented elsewhere in the East African rift valley (Baker and others, 1972) and indeed may be characteristic of major rift valleys (Withjack, 1979). However, it should be noted that the scale of the doming in the Eyasi region appears to be much smaller than that observed or inferred elsewhere in association with major rift-valley faulting, as for example in the Kenya sector of the eastern rift valley (Baker and others, 1972).

Conclusions

1. The Laetolil Beds were deposited 3.5 to 3.7 m.y. B.P., on the crest and flanks of a broad domal uplift. Rift-valley faulting began about 2.1 m.y. B.P. and continued into the last Pleistocene, uplifting the Eyasi Plateau and the Laetolil Beds to their present elevation.

2. The upper, fossiliferous part of the Laetolil Beds is chiefly wind-worked and air-fall tuffs of nephelinite, melilitite, and carbonatite composition. These tuffs were weathered before burial and can be considered a weakly developed sequence of caliche-type paleosols.

3. The faunal assemblage is similar in most respects to that of the present time and points to a grassland savanna lacking major rivers or nearby lakes.

4. Both lithologic and faunal evidence indicates that the climate was semiarid, and the Footprint Tuff provides evidence of well-defined dry and rainy seasons. The climate was probably somewhat warmer than at present.

Acknowledgments

Study of the Laetolil Beds was supported by the National Science Foundation (Grants EAR 71-01523 and 76-84583) and the National Geographic Society. I am indebted to A. Mturi for permission of the Tanzanian government to work at Laetoli and to M. D. Leakey for use of a vehicle and camp facilities. M. Monaghan, R. J. Clarke, and John Kang'wezi assisted in the field work. The drafting was done by Judith Ogden.

References

Baker, B.H., Mohr, P.A., and Williams, L.A.J., 1972, Geology of the Eastern Rift System of Africa: Geological Society of America Special Paper 136, 67 p.

Cooke, H. B. S., and Maglio, V.J., 1972, Plio-Pleistocene Stratigraphy in East Africa in relation to proboscidean and suid evolution; in Calibration of Hominid Evolution: eds., Bishop, W.W., and Miller, J.A., pp. 303-329. Edinburgh, Scottish Academic Press, 486 p.

Dietrich, W.O., 1942, Altestquartare saugetiere aus der sudlichen Serengeti, Deutsch-Ostafrica: Palaeontographica, v. 94, p. 43-133.

Hay, R.L., 1976, Geology of the Olduvai Gorge, Berkeley, University of California Press, 203 p.

_____, 1978, Melilitite-carbonatite tuffs in the Laetolil Beds of Tanzania: Contributions to Mineralogy and Petrology, v. 67, p. 357-367.

Hay, R.L., and Reeder, R.J., 1978, Calcretes of Olduvai Gorge and the Ndolanya Beds of northern Tanzania: Sedimentology, v. 25, p. 649-673.

Hooijer, D.A., 1979, Hipparions of the Laetolil Beds, Tanzania: Zoologische Mededelingen, v. 54, p. 15-33.

Johanson, D.C., and White, T.D., 1979, A systematic assessment of early African hominids: Science, v. 203, p. 321-330.

Kent, P.E., 1941, The recent history and Pleistocene deposits of the plateau north of Lake Eyasi, Tanganyika: Geological Magazine, v. 78, p. 173-184.

Kohl-Larsen, L., 1943, Auf den Spuren des Vormenschen; Forschungen Fahrten und Erlebnisse in Deutsch-Ostafrika (Vol. 2): Stuttgart, Strecker and Schroder Verlag, 394 p.

Leakey, M.D., Hay, R.L., Curtis, G.H., Drake, R.E., Jackes, M.K., and White, T.D., 1976, Fossil hominids from the Laetolil Beds: Nature, v. 262, p. 460-466.

Leakey, M.D., and Hay, R.L., 1979, Pliocene footprints in the Laetolil Beds at Laetoli, northern Tanzania: Nature, v. 278, p. 317-323.

Pickering, R.P., 1964, Endulen: Geological Survey of Tanzania Quarter degree sheet 52, scale 1:125,000.

Withjack, M., 1979, An analytical model of continental rift fault patterns: Tectonophysics, v. 59, p. 59-81.

2. Lithofacies and Environments of Bed I, Olduvai Gorge, Tanzania

Introduction

Approach to Environmental Analysis

The present environmental analysis of Bed I is based on its subdivision into lithologic facies, which are lithologically different but laterally equivalent rock assemblages, each of which was deposited in the same environment or a closely related series of environments. The facies are then analyzed for evidence of (1) paleogeography and physical sedimentation processes, (2) chemical environment of sedimentation and postdepositional processes (e.g., weathering), (3) rate of deposition and time represented by paleosols, (4) nature of the biota, and (5) climate. Data from each facies are then integrated into an environmental synthesis and geologic history for the entire basin. Selley (1970) elegantly demonstrates the use of lithofacies in analyzing stratigraphic units in terms of depositional environment.

Several features make Bed I particularly well-suited for environmental analysis. It was deposited in a small basin, easy to study, and encompasses several different lithologic facies. Source areas for detrital sediment vary considerably in their composition, and thus the detrital sediments of Bed I can generally be assigned to specific sources. With the sediment sources known, stream-channel alignments can be used to determine the drainage pattern. Several tuffs are widespread, and the paleogeography at the time of any major ash fall can be readily inferred by examining the lithology and thickness of the resulting tuff at different places. Chemically precipitated minerals in the lake deposits provide a wealth of information about the chemistry of the lake

This paper originally appeared in Quaternary Research (v. 3, pp. 541–560; © University of Washington); reprinted by permission.

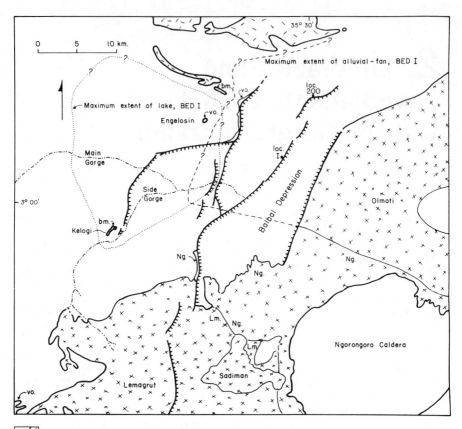

BASEMENT METAMORPHIC ROCKS

VOLCANIC ROCKS

FAULT

Figure 1. Map of the Olduvai region showing principal faults, volcanoes, and outcroppings of metamorphic basement rock. Abbreviations used for volcanoes are Ng = Ngorongoro and Lm = Lemagrut. Dotted line outlines the inferred maximum extent of the Bed I lake; dashed line indicates the maximum extent of the alluvial fan of Bed I.

water. Finally, the Olduvai fauna is rich, varied, and well-
studied, and several of the faunal elements have proved to be
diagnostic of specific environmental conditions.   Limited
exposure of Bed I outside of the gorge is the principal ob-
stacle to a fully satisfactory paleogeographic reconstruction.

Geologic and Climatic Setting

Olduvai Gorge is a steep-sided valley in the Serengeti
Plain near the western margin of the Eastern Rift Valley in
northern Tanzania (Fig. 1). Where cut by the gorge, the
plain has an elevation of 1350 to 1500 m. The gorge is
generally 45 to 90 m deep in the lower 26 km of its course,
where Bed I is exposed. About 9 km upstream from its mouth
the gorge divides into two branches--a smaller, southern
branch, the Side Gorge, and a larger, northern branch, the
Main Gorge. Rainfall varies widely but averages 65 cm/yr
over the period 1966-1971. The temperature ranges from 14°
to $32^{\circ}C$, averaging about $22^{\circ}C$ over parts of the year 1972 in
which records were kept. The Serengeti Plain is a grassland
with Commiphora scrub and scattered Acacia (thorn trees).

Metamorphic rocks of Precambrian age, principally
gneisses and quartzites, are exposed in the upstream part of
the gorge, in inselbergs in the vicinity of the gorge, and in
highlands to the north. A trachytic welded tuff, the Naabi
Ignimbrite, overlies the metamorphic basement in the western
part of the gorge. It was probably erupted from Ngorongoro,
which contains similar rocks. The Naabi Ignimbrite is
overlain by the Olduvai Beds: Beds I through IV, the Masek
Beds, the Ndutu Beds, and the Naisiusiu Beds (Hay, 1971).
Beds I through IV accumulated in a small, shallow basin along
the northwestern foot of the volcanic highlands as represented by Ngorongoro, Lemagrut, and Olmoti.

All but the uppermost unit of the Olduvai sequence is
cut by numerous faults. Reck (1951) numbered the larger of
the faults from east to west as the First, Second, Third,
Fourth, and Fifth faults. Faulting resulted in the Balbal
Depression, a graben into which the gorge presently drains.
The faulting began early during the deposition of Bed II and
continued intermittently through deposition of the Ndutu
Beds.

Bed I is the lowest unit of the Olduvai Beds, and where
fully exposed in the western part of the gorge it is
generally 30 to 43 m thick. The only complete section in the
eastern part of the gorge, near the Third Fault, is 54 m.
Here Bed I is underlain by a semiwelded tuff, or ignimbrite,
probably erupted from Ngorongoro. This ignimbrite differs

mineralogically from the Naabi Ignimbrite, and the two are not correlative. By comparison with the Ngorongoro Ignimbrite sequence, that at the Third Fault should be younger than the Naabi Ignimbrite, and very possibly it interfingers westward into the lower part of Bed I.

The southern and western margins of the basin of Bed I can be located in a few places with a reasonable degree of certainty. Bed I pinches out against basement rocks and the Naabi Ignimbrite in the western part of the gorge. To the southwest, near Kelogi inselberg, it thins to a few meters and very likely pinches out over massive nephelinite tuffs correlated with the Laetolil Beds. Bed I apparently pinches out several kilometers south of the mouth of the gorge.

Bed I extends at least 15 km, and probably much more, to the northeast of the gorge. Lavas of Bed I are exposed in fault scarps 5 to 12 km north of the gorge, and tuffs and claystones crop out along the east-west fault that parallels the gorge 5 km to the north (locs. 201, 202). The most distant exposure is 15 km to the northeast (loc. 200), where the uppermost 7 m of Bed I can be seen. Hills of basement rock 12 km north of the gorge very likely form the northern boundary of the Bed I Olduvai lake basin, although lacustrine deposits of the same age may possibly underlie the plains area separating these hills from the highlands 3 to 5 km farther north. As inferred from outcrop distribution and lithologic patterns, the Olduvai lake basin at the time of Bed I may have had maximum north-south and east-west diameters of roughly 25 km (Fig. 1).

Tuffs provide the principal basis for correlating in Bed I (Hay, 1971). The six tuffs most useful in correlating have been designated Tuff IA, IB, IC, ID, IE, and IF (Hay, 1971). Tuffs IA, IC, ID, and IF are ash-fall tuffs. Tuff IB comprises an ash-flow tuff, both primary and reworked, and a widespread ash-fall tuff. Tuff IE is an ash-flow tuff. The oldest of the artifacts and hominid remains lie a short distance beneath Tuff IB; the others lie at varying levels between Tuffs IB and IF.

Three geophysical dating techniques have been applied to Bed I, giving at least the upper part of it one of the firmest dates of any fossiliferous lower Pleistocene stratigraphic unit. Fifty-seven K/Ar dates have been obtained from Bed I, clearly indicating an age on the order of 1.7 to 1.8 m.y. for the fossiliferous deposits. The most satisfactory dates are from Tuff IB, which has an age of $1.79 \pm .03$ m.y. (Curtis and Hay, 1972). The K/Ar dates receive supporting evidence from a fission-tract date of $2.03 \pm .28$ m.y. (Fleischer, et al., 1965). More recently, the magnetic stratigraphy has been worked out, at least roughly, for the Olduvai Beds (Fig. 2), and this gives a more satisfactory basis for estimating the duration of Bed I than do the K/Ar dates. Briefly, a normal polarity event, the Olduvai event (~1.86 to 1.71 m.y.), is recorded by Bed I and the lower part of Bed II, including the lowermost part of the Lemuta Member (Gromme and Hay, 1971). Both older and younger rocks have reversed polarity. On the basis of strata thickness and development of paleosols, I estimate that the fossiliferous, artifact-bearing part of Bed I represents a time span of 50,000 to 100,000 yr. The uppermost ignimbrites of Ngorongoro have given K/Ar dates of 2.0 ± .1 (Curtis and Hay, 1972) and 2.1 ± .1 m.y. (G.H. Curtis in Brock et al., 1972). The lowermost deposits of Bed I in the western part of the gorge are probably of the same age but may be slightly younger, depending on whether or not the youngest ignimbrite of Ngorongoro interfingers with Bed I.

Lithologic Facies

Bed I can be subdivided into five lithologically different but laterally equivalent rock assemblages, or lithologic facies, which are the primary basis for interpreting Bed I in terms of environment. Four of the facies are named for the environment in which they were deposited: lake deposits, lake-margin deposits, alluvial-fan deposits and alluvial-plain deposits. Lava flows constitute the fifth lithologic facies. It should be emphasized that these facies names oversimplify the environmental interpretation, as some of the facies contain sediments deposited in two or more environments.

Facies relationships are shown in Figure 3 which is an east-west cross section through Bed I at the end of its deposition. The lake deposits accumulated in a perennial lake in the lowest part of the basin. Lake-margin sediments were laid down on low-lying terrain intermittently flooded by the lake. Lava flows and alluvial-fan deposits are along the eastern margin of the basin and they interfinger westward with lake-margin deposits. Alluvial-plain deposits are exposed only in the western half of the basin, where they underlie lake and lake-margin deposits. Lake-margin deposits of the eastern and western parts of the basin differ in lithology as well as fossils and archaeologic content, and the two assemblages of lake-margin sediments will be described separately under the headings Eastern lake-margin deposits and Western lake-margin deposits. The lake-margin sediments to the north and south of the gorge are included in the eastern lake-margin deposits on the basis of lithology.

The following sections will describe the lithologic

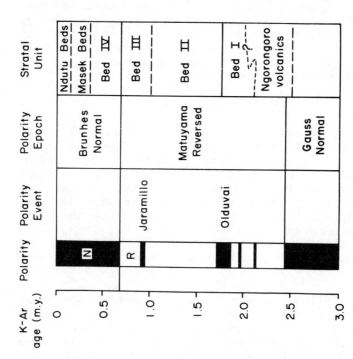

Figure 2.   Diagram showing ages of Ngorongoro and the Olduvai
Beds as inferred from magnetic stratigraphy and K/
Ar dates.   The top of the Matuyama Reversed Epoch
may lie higher than the base of Bed IV, as only a
few polarity measurements, by A. Brock, are presently available for the lower part of Bed IV.

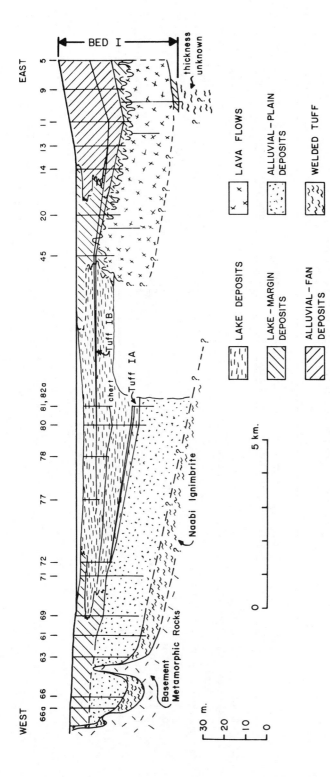

Figure 3. East-west cross-section of Bed I, showing its lithofacies as exposed in the Main Gorge. Numbers refer to localities of measured sections, and vertical lines indicate thickness of section measured (see Fig. 4: also location map in Leakey, 1971). Cross-section is reconstructed with the top of Bed I approximately as it would have appeared when Tuff IF was deposited.

Table 1. Composition of Lithofacies in Bed I

Facies	Percent Claystone	Percent Sandstone	Percent Conglom- erate	Percent Tuff	Limestone and dolomite	Other
Lake deposits	80	2	tr.	15	3	tr. chert
Lake-margin deposits (eastern)	49	tr.	tr.	49	2	tr. siliceous earth
Lake-margin deposits (western)	40	42	.3	14	4	
Alluvial-fan deposits	2	0	2	96[a]	0	
Alluvial-plain deposits	28	4	3	56	9	
Lava flows	0	0	0	tr.	0	>99% lava flows

[a] Includes about 10% of pyroclastic deposits coarser than tuffs.

facies of Bed I and interpret them in terms of depositional environment. All descriptions are extremely brief except for the eastern lake-margin deposits, which contain all of the known hominid fossils and archaeologic materials. The lake deposits are described first as the nature of the lake is basic to an understanding of the lake-margin deposits. Facies descriptions will then be in the following sequence: western lake margin, eastern lake margin, alluvial fan, lava flows, and alluvial plain.

Lake Deposits

The lake deposits are exposed westward about 6.5 km from the Fifth Fault and extend an undetermined distance to the east beneath younger sediments. They have a maximum thickness of 26 m near the Fifth Fault (e.g., loc. 80), and thin to about 10 m near the western margin. Lake deposits interfinger and intergrade westward with lake-margin deposits over a zone about 2 km wide. The lake deposits are chiefly claystone but include tuff, sandstone, limestone, dolomite, chert, and conglomerate (Table 1).

Claystones are generally wax-like, massive to laminated, and pale olive to greenish-gray. Authigenic minerals, that is, minerals formed in place, constitute a substantial amount of the claystones. Widespread and abundant are calcite, dolomite, K-feldspar, and altered pyrite; relatively rare are calcite replacements of gaylussite and trona(?). Calcite crystals constitute, on the average, between 5 and 10% of the claystones. They are unusual in their large size (generally .2 to 2 mm long), euhedral shapes, and varied crystal habits at different levels. Calcareous claystones grade into clayey limestones consisting largely of coarse calcite crystals. A single horizon of desiccation cracks was found near the middle of the facies in one of its easternmost exposures.

Tuff beds are laterally extensive, evenly bedded, and decrease westward in both grain size and thickness. Most of them are vitric tuffs of original trachyte composition in which the glass has been altered to phillipsite and K-feldspar.

Limestones are widespread, and dolomites are found in the eastern half of the facies. One limestone is oolitic, but the others are of coarse calcite crystals similar to those of the claystones. Dolomites are characteristically very fine-grained. Chert nodules are both widespread and abundant in a single horizon near the middle of the facies. Sandstones and pebble conglomerates are present in a few places in the western part of the facies.

These sediments accumulated in a shallow saline lake of fluctuating level and extent. The greenish pyritic clays are a typical lacustrine sediment, and the laminations in clays and even-bedding of tuffs and dolomites are evidence of quiet waters. Evidence for fluctuation is provided by the variable western margin of the facies and interbeds of pebbly sandstone within wax-like claystones. Desiccation cracks indicate exposure to the air, and sand-size claystone pellets at a few localities reflect erosion of nearby mudflats, probably by wind (Price, 1963). Casts of the soluble sodium-carbonate minerals gaylussite and trona(?) are the most direct proof of a highly saline sodium-carbonate lake, but the authigenic dolomite and K-feldspar are strongly suggestive (Hay, 1966, 1970). The chert nodules are of a type found only in deposits of saline sodium-carbonate lakes (Hay, 1968).

The coarse, euhedral calcite crystals deserve special comment, as only a single other occurrence of this type has been reported, as far as I am aware (Isaac, 1967). At least some of the crystals grew in lake-bottom muds at depths sufficiently shallow (a few centimeters or less?) for wave action to rework the muds and concentrate the crystals in thin layers. Isotope measurements were obtained by J. R. O'Neil from 12 samples of calcite crystals in order to estimate the salinity of the fluid from which the calcite was crystalized. This salinity determination is based on the principle that the heavier isotopes are selectively concentrated (e.g., ^{18}O over ^{16}O) in the evaporation required to form a brine from dilute meteoric water. Both the $^{18}O/^{16}O$ and $^{13}C/^{12}C$ ratios are high in all samples: $\delta^{18}O$ = +28.5 to 34.5, averaging +30.6‰ (per mil); $\delta^{13}C$ = +4.7 to 7.9, averaging to +6.1‰. These values are much too high for freshwater calcite (see Graf, 1960) but are compatible with crystallization from a brine. Very likely the calcite crystals were precipitated in the dissolution of gaylussite ($Na_2Ca(CO_3)_2 \cdot 5H_2O$), which would supply both Ca^{2+} and CO_3^{2-} for growth of calcite in lake-bottom muds. Oxygen-isotope measurements on a chert nodule ($\delta^{18}O$ = +36.4‰) also point to saline lake water (O'Neil and Hay, 1973).

Western Lake-Margin Deposits

The western lake-margin deposits are 8 to 12 m thick and extend westward 3 km from the zone of interfingering with the lake deposits. These sediments are chiefly sandstone (42%) and claystone (40%) but include a substantial amount (14%) of tuff. Sandstones are chiefly quartz and feldspar of basement origin. Oolites are in many of the easternmost sandstones, and claystone pellets are abundant in some of those to the west.   The conglomerates consist of clasts of the Naabi Ignimbrite and basement rock.   Claystones are grayish-brown or pale olive and contain rootmarkings.   Some are sandy, and the sandier of the claystones contain claystone pellets. Trachyte tuffs are in the upper 4.5 m of the facies, and mafic (basaltic?) tuffs are in the underlying part.   Reworked mafic vitric tuff and tuffaceous sandstone of Tuff IA locally form the basal bed of the facies, and an evenly laminated mafic vitric tuff 2 m thick lies near the middle of the facies on the south side of the gorge.   The trachytic tuffs above correlate with those of the sequence from Tuff IB to IF in the lake deposits to the east.   They are thin-bedded in their eastern exposures and massive and rootmarked to the west.   Rounded pumice cobbles as much as 15 cm in diameter were noted in Tuff IF at locality 64.   There are a few steep-sided stream channels, filled with tuff or conglomerate.   Two channels near the base of the facies are oriented N65° E and N85° E; two others near the top are N50° E and N80° E.   The four channels average N70° E.   The deepest channel, at locality 60, has a depth of 3.4 m.   As the detritus in this facies is from westerly sources, the channels were cut by streams flowing northeast.

An antelope skull was found in locality 63 either at the base of the lake-margin deposits or in the topmost alluvial-plain deposits.

These deposits have both lacustrine and fluviatile features, indicating that they accumulated on terrain inter-mittently flooded by the lake.   Oolitic sandstone indicates shallow, wave-agitated lake water, and the laminated vitric tuff suggests quiet water.   Sand-free claystones were deposited in quiet water, either of a lake or flood plain. Lenticular conglomerates and channeling are fluvial features, and if the streams flowed into the lake perpendicular to its margin, then the shoreline was oriented about N20° W.   The stream channel 3.4 m deep was near the margin of the lake, and its depth may be a rough measure of the fluctuation in lake level.   Claystone pellets probably reflect erosion of mudflats by wind at a time of low lake level.   In view of their large size, the rounded pumice cobbles probably floated westward across the lake to their present position.

The lake water flooding the lake-margin terrain was overall less saline than that in the center of the basin. Vitric tuffs are extensively altered to zeolites only in the eastern part of the facies, and both authigenic K-feldspar and evidence of soluble salts are lacking.

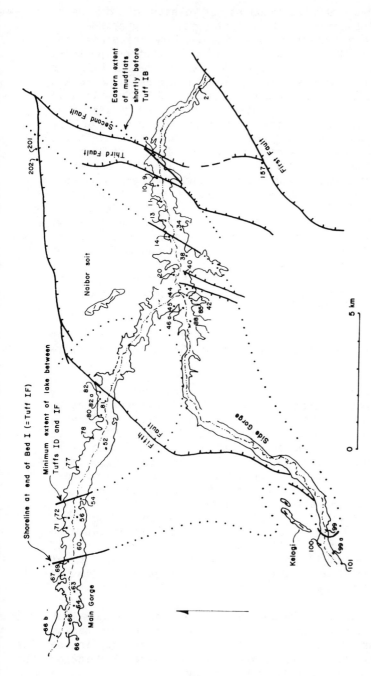

Figure 4. Map showing outline of Olduvai Gorge, principal faults, and localities referred to in text. Also indicated are the shoreline at the end of Bed I (=Tuff IF), the minimum extent of the lake between Tuffs ID and IF, and the eastern extent of lake-margin mudflats shortly prior to eruption of Tuff IB. These paleogeographic features are shown as solid lines where based on exposures in the gorge and as dotted lines where inferred from overall paleogeographic considerations. Localities 11 (= MK), 13 (= DK), and 45 (= FLK) are sites of excavations for hominid remains and archeologic materials. Not indicated on map are localities 45a (= FLK-N) and 45b (= FLK-NN), which lie, respectively, about 70 and 180 m northwest of FLK (see Leakey, 1971).

Eastern Lake-Margin Deposits

Lake-margin deposits to the east of the lake beds are exposed almost continuously in the Main Gorge above the lavas over an east-west distance of 5.4 km. These are as much as 17 m thick, and an additional 1.4-m thickness is present beneath the lavas. Also included in these deposits are 5 m of beds near Kelogi, to the southwest (locs. 99, 100), and 7 m of beds to the north of the gorge (locs. 201, 202).

The vertical sequence of lake-margin deposits above the lavas is subdivided at the base of Tuff IB into lower and upper units. The lower unit extends from the Second Fault west to FLK-NN, a distance of 5.4 km (Figs. 4, 5). The upper unit is restricted by alluvial-fan deposits to the western half of this area. The lower unit, except for its easternmost exposures, is 0 to 6 m thick, the variation reflecting the uneven surface of the lavas. It consists largely of claystone but contains a few trachytic tuffs. Granule- and pebble-size clasts of basement and volcanic rock occur widely on the north side of the gorge. These sediments grade eastward into a section that is dominantly tuff and is as much as 11 m thick. These easternmost deposits have some lithological similarity to the overlying alluvial-fan deposits and some mineralogical similarity to the underlying lavas. They are, however, included within the lake-margin deposits in view of their limited exposure and uncertain genetic relationship. The upper unit of lake-margin deposits is about two-thirds tuff and one-third claystone, and it also includes limestone, pumice-pebble conglomerates, and siliceous earth.

Overall, the eastern lake-margin deposits are about 98% tuff and claystone, which are in roughly equal proportions. There is perhaps 2% of limestone and far less than a percent each of conglomerate and siliceous earth. Paleosols are weakly developed at many horizons over both tuffs and claystones. Where developed on claystones, they are crumbly, brown or brownish-gray, rootmarked zones. Paleosols on tuffs are rootmarked zones as much as 30 cm thick in which much of the finest-grained vitric ash is weathered to montmorillonite.

Lithologic descriptions. Claystones are both tuffaceous and nontuffaceous and can be pale olive or various shades of yellow and brown. Rootmarkings are common particularly above the level of Tuff IC. Wax-like claystones, relatively free of sand, are much more common below Tuff IC than they are above. Calcite concretions are widespread and abundant, and calcite replacements of gypsum rosettes ("desert roses") been found beneath Tuffs IB and IF.

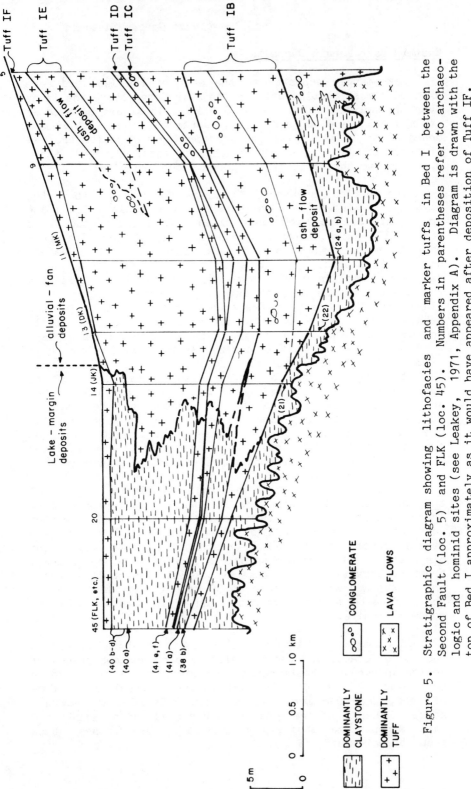

Figure 5. Stratigraphic diagram showing lithofacies and marker tuffs in Bed I between the Second Fault (loc. 5) and FLK (loc. 45). Numbers in parentheses refer to archaeologic and hominid sites (see Leakey, 1971, Appendix A). Diagram is drawn with the top of Bed I approximately as it would have appeared after deposition of Tuff IF.

Tuffs are dominantly trachytic but include some of basaltic and trachyandesitic composition. Basaltic tuffs are confined to the eastern margin of the facies, where they overlie the lava flows. The other, more widespread tuffs vary considerably in the degree and nature of reworking and provide considerable information about the environment in which they were deposited. A few tuffs are massive, root-marked, and represent single showers of ash that fell on the land surface (e.g., Tuff IB at FLK-NN). A massive 1.2-m-thick bed above Tuff ID is rootmarked and contains diatoms and ostracods in addition to pyroclastic materials. It is a product of several eruptions in which the ash was deposited either in marshland or on a land surface intermittently flooded by lake water. A few tuffs are for the most part evenly laminated and are dominantly lacustrine. Tuff IF is a lacustrine deposit, and its thickness ranges from 15 cm to 2.1 m and varies systematically on a regional scale, suggesting that the greater thicknesses accumulated in deeper water. Tuff IF is also laminated and waterlaid at localities 201 and 202, to the north of the gorge, and locality 99 to the southwest. This tuff is zeolitic in its northwestern exposures (e.g., locs. 88, 45) and unaltered to the southeast (e.g., locs. 38, 40). Limestones are almost entirely of two types -- either coarsely crystalline concretions cemented together or fine-grained beds of irregular shape. Both of them occur within claystones and were probably precipitated from ground water in clay at shallow depth. Isotopic values from two limestone concretions are much lower than those on calcite crystals of the lake deposits. One concretion, from claystone beneath the "Zinjanthropus" level, gave $\delta^{18}O$ = +24.3‰ and $\delta^{13}C$ = 5.7‰. Another, from 1 to 2 m below Tuff IF at FLK-N, gave $\delta^{18}O$ = +24.1‰; $\delta^{13}C$ = -5.8‰. (J.R. O'Neil, personal communication). These values are compatible with crystallization from relatively freshwater (Graf, 1960).

Siliceous earth is used here to refer to a white or cream-colored silica-rich deposit that is friable, porous, and has an earthy luster. Silica is in the form of fine-grained particles of biogenic opal. A 60-cm thickness of siliceous earth was found at one place in the widespread rootmarked massive tuff that overlies Tuff ID. In a few places elsewhere the tuff is siliceous and earthy. The top-most 3 cm of claystone beneath Tuff IF at locality 46a has laminae of siliceous earth.

An extraordinary variety of materials makes up the assemblage of granule (2 to 4 mm) and pebble-size (4 to 64 mm) clasts in the lower unit between localities 10 and 20. The clasts comprise basement and volcanic detritus, commonly accompanied by fossilized bone fragments. Clasts may either be isolated or concentrated in thin, lenticular beds. The basement clasts are quite varied and include materials (e.g., pink pegmatitic feldspar) from an unknown source, probably to the north or northwest. Clasts of volcanic rock are chiefly welded tuff from Ngorongoro but include varied lavas, a little of which is from Engelosin, an eroded volcanic neck of distinctive nepheline phonolite which lies 8 km north of the gorge. Conglomerates elsewhere in the lower unit are chiefly of lava and welded tuff from Ngorongoro. Pumice-pebble conglomerates predominate above the level of Tuff IB.

Flora and fauna. Fossilized leaves are rare, and little is known of the pollen, but swamp vegetation is indicated by abundant coarse, generally unbranching, vertical root channels and casts. These are commonly 10 to 30 cm long, 2 to 6 mm wide, and could have been made by Typha and some kinds of reeds. Fossil rhizomes of papyrus (?) point to marshland or shallow water. Silicified remains of a water plant (cf. Potamageton) were identified by Howard Schorn (1971, personal communication). Diatoms are common in the more siliceous parts of the 1.2-m thick tuff unit above Tuff ID (esp. loc. 85). Dr. Jonathan Richardson identified eight genera in one sample, and his comments are as follows (1972, personal communication):

"The numbers involved were insufficient to be very sure that the diatoms were deposited in situ. They are mainly benthic and epiphytic types, rather than planktonic (only the Melosira is probably planktonic, and I found only one specimen of that). Surirella fasciculata and Rhopalodia gibberula, which are perhaps the most characteristic types here, are both typical of fairly alkaline water. I would hazard the speculation that this was a small and impermanent water body in which the soil remained moist enough during the dry spells to sustain sedges and other semi-aquatic types of plants."

Generally associated with diatoms are hollow, knobby spheres of opaline silica 7 to 15 µm in diameter that Howard Schorn identifies as encysting cases of chrysophyte algae which point to fluctuating lake levels and periodically saline conditions.

Ostracods, indicative of quiet water, are in a few samples of the siliceous tuff unit above Tuff ID, and freshwater snails were collected from clays of the "Zinjanthropus" level (Leakey, 1971). Fossil remains of urocyclid slugs have been found at two levels below Tuff IC (ibid.). These suggest damp conditions in evergreen forests "where the rainfall exceeds 35 inches per year or where damp conditions are maintained by regular mists" (Verdcourt, 1963).

Crocodile remains are widespread below Tuff IB and are especially common at DK (loc. 13). Remains of fish were collected from the "Zinjanthropus" level at FLK (loc. 45) and from the uppermost part of Bed I at FLK-N. Regarding the fish, Greenwood and Todd (1970, p. 241) state:

"Site FLK-NI is of particular interest because it has yielded only the remains of Cichlidae. Samples from this site are quite small, but, by analogy with other sites, at least a few clariid remains would be expected. Clariids are some of the most ubiquitous fishes in present day African fresh waters. Their absence, when cichlids are present, might indicate strongly saline (especially alkaline) conditions, as in Lake Magadi today. Since earlier and later fishbearing deposits contain both clariids and cichlids, there does seem to be a possibility that the lake was more saline during the deposition of the FLK-NI bed than at other periods."

The fossil avifauna of Beds I and II is one of the largest heretofore known. Dr. P. Brodkorb has summarized the general nature of the avifauna as follows (1973, personal communication):

"Abundant remains of aquatic birds indicate that water was much more plentiful throughout the time of deposition of the earlier strata that form Beds I and II than is the case under the present semidesert conditions. The water birds included swimmers and divers such as grebes, cormorants, pelicans, and many ducks. Larine birds were represented by gulls, terns, and skimmers. Waders were abundant and included flamingoes, herons, storks, rails, jacanas, plovers, sandpipers, and stilts. The presence of flamingoes at several sites in Both Bed I and Bed II indicates the proximity of brackish water. Seed-eaters include francolins, quail, hemipodes, and several species of doves. They demonstrate the presence of grassland."

Aquatic birds are most common below Tuff IC, and land birds (perching birds, quail, parrots, etc.) are most common in the uppermost levels at FLK-N.

The mammalian fauna is likewise varied and abundant. At many levels are remains of relatively large mammals: proboscideans (principally elephants), rhinoceros, equids, suids, hippos, bovids, and giraffids. Dr. M.D. Leakey (1971) gives the number and proportions of specimens of identifiable larger mammals from excavated sites. In this tabulation the bovid remains, principally of gazelles and other antelopes, are much more abundant between the "Zinjanthropus" level and Tuff IF than they are in the underlying deposits. This difference may possibly reflect the increasing importance of bovids in the hominid diet rather than any real difference in the fauna.

Remains collected from the 1.2 m of claystone below Tuff IF at FLK-N (loc. 45a) are of strikingly different character. The fauna is especially rich in small mammals, as for example, rodents, gerbils, Heterocephalus ("naked rat"), and other rodents, and carnivores, especially viverrids (e.g., genets, civets). This fauna indicates significantly drier conditions than that collected from lower levels. Of the rodents, Lavocat has stated (in L.S.B. Leakey, 1965, p. 19), "A certain number of forms in this fauna from FLK-NI--and they are the more numerous, moreover--seem to indicate a steppe, or subdesertic environment."

To summarize, both floral and faunal evidence points to widespread marshland and the frequent occurrence of standing water up through deposition of the siliceous earthy tuff above Tuff ID. Drier conditions are strongly suggested by fish, birds, and rodents form excavated levels in the uppermost 1.2 m below Tuff IF at FLK-N.

Depositional environment. A wealth of evidence shows that these deposits accumulated on relatively flat terrain that was intermittently flooded and dried in response to changes in level of the lake. Prior to Tuff IB, the zone of marginal terrain along the eastern margin of the lake was at least 5.4 km (Fig. 5); after Tuff IB, the zone was narrowed by about 3.4 km in westward growth of an alluvial fan.

Claystones and widespread water-laid tuffs were deposited at times of high water level, and paleosols, homminid occupation sites, eroded surfaces, and extensive land-laid tuffs represent periods of exposure.

Most of the fluctuations may have been relatively short--either seasonal or involving no more than a few tens of years. A longer-term paleogeographic or climatic fluctuation is indicated by the fauna collected at several horizons through 1.2 m of claystone beneath Tuff IF at FLK-N.

This fauna points to substantially drier conditions than prevailed at lower horizons for which the fauna is known. The 2.4 m of deposits beneath the lavas indicate a lake-margin terrain similar to that between the lavas and Tuff IB.

The pattern of zeolitic alteration in lacustrine tuffs points to a salinity gradient at times of flooding, with the greater degree of zeolitic alteration and therefore the highest salinities on the northwest and the freshest water on the southeast, nearest the stream inlets. This gradient is shown most clearly in Tuff IF, which is lacustrine over the entire width of the lake-margin terrain.

Prior to eruption of Tuff IB, streams from Ngorongoro carried pebbles of lava and welded tuff westward over the mudflats. It is not clear, however, as to how the granules and pebbles of basement rock (quartz, pink feldspar, and gneiss) and Engelosin phonolite were transported southward and southeastward for a considerable distance. Hominid activity can be eliminated as a general explanation in view of the large number and broad distribution of the clasts. The clasts are in claystones rather than in beach deposits, and sheetwash associated with torrential downpours to the north of the gorge seems the most likely way of transporting coarse sediment southward over mudflats.

Initially, the lava surface had a local relief of at least 6 m, which was reduced as the lavas were buried by sediments. The lake-margin terrain had a relief of about 2 m just prior to Tuff IF, which filled in topographic irregularities.

Alluvial-Fan Deposits

Alluvial-fan deposits lie along the east side of the basin and interfinger westward with lake-margin deposits. They are exposed in the gorge over an east-west distance of 7.5 km, and they crop out intermittently for a distance of 5 km to the northeast along the west side of the Balbal depression. The uppermost part of the facies is exposed 15 km northeast of the gorge (loc. 200).

The alluvial-fan deposits are almost entirely of eruptive origin and they range from 21 to 28 m in thickness. Tuff IB is the basal bed of these deposits. Approximately 15% of the facies consists of ash-fall and ash-flow deposits that have not been reworked. Ash-flow deposits exposed in the gorge comprise Tuff IE and the massive part of Tuff IB that overlies a basal ash-fall stratum. Their magnetic polarity shows that they were emplaced at high temperature

(Grommé and Hay, 1971). The reworked pyroclastics were deposited chiefly by streams but include some layers emplaced as mud flows. A few of the fine-grained tuffs are thinly laminated. Overall, the alluvial-fan deposits are about 90% glass and 10% crystals and rock fragments. Nearly all are of sodic trachyte composition, and a very few are probably trachyandesite.

Brown, rootmarked paleosols, 15 to 60 cm thick, are widely developed on Tuffs IB and IC and on tuffs that underlie Tuffs IC and IF. More localized and poorly defined paleosols are developed at a few other horizons above Tuff IC in the vicinity of the Second and Third Faults, and rootmarkings are through most of the section near the western margin of the facies. Tuffs are generally unstratified in the paleosols developed on thinly-bedded tuffs, showing the churning effect of roots and burrowing animals. The paleosols have weakly developed profiles, and only a small amount of clay has been formed by weathering.

Conglomerates form lenticular beds that are coarsest and most abundant in the eastern part of the facies. Clasts are chiefly of cobble and pebble size, but boulders as much as 60 cm in diameter are not rare. The clasts are of pumice, obsidian, lava, welded tuff, and zeolitic lapilli tuff.

Seventeen stream channels were measured between locality 34 and the Second Fault. Nearly all of them are 1 to 1.6 m deep and filled with conglomerate. Eight of these channels are cut into the ash-flow deposit of Tuff IB; the others lie at various horizons above. Their orientations range from about east-west to N35° W, averaging N73° W. Eleven of these, including all of the deeper and better-defined channels, are between N60° W and N85° W.

Remains of two vertebrates have been collected from alluvial-fan deposits at DK (loc. 13). One is a suid, represented by a skull and mandible, from between Tuffs IC and ID. The other is *Parapapio* sp., a skull of which was found on the surface (M.D. Leakey, 1971, personal communication). No other faunal remains have been found.

These deposits accumulated on the lower part of an alluvial fan, sloping northwestward from Olmoti and Ngorongoro to the lake-margin terrain. The relatively consistent stream-channel alignments and coarse size of much debris point to significant slopes, although some of the laminated sediments indicate local areas of ponded water. An average slope of 1:300 (3.3 m/km) is used in the cross sections of Figures 3 and 5. This slope seems reasonable on the basis of alluvial fans in the San Joaquin Valley of California.

Vegetation grew periodically on the fan as shown by the rootmarked paleosols, but its surface must have been relatively barren for much of the time. The scarcity of mammal remains probably reflects the sparseness of the vegetation. Both the sparse vegetation and mudflow deposits are compatible with a semiarid climate as indicated by evidence in other facies.

Pyroclastic materials of the fan were erupted from Olmoti. Average clast size is coarsest, and ash-flow tuffs are most numerous in locality 1, the exposure closest to Olmoti. The average stream channel alignment (N73° W), when extrapolated upslope to the southeast, touches the volcanic highlands near the junction of Ngorongoro and Olmoti. As the central crater of Ngorongoro had been extinct long before this time, Olmoti is indicated as the source. Most of the lava and tuff clasts in conglomerates fit best with a source on Olmoti.

Lava Flows

Lava flows are exposed almost continuously in the gorge from the Second Fault west to FLK-NN, a distance of 5.4 km. They have a total thickness of 20 m where they are fully exposed upstream from the Third Fault. Lavas of Bed I also crop out along the western margin of the Balbal Depression, and 45 to 60 m of flows are exposed in fault scarps 5 to 12 km north of the gorge. In the gorge near the Second and Third Faults, the lava sequence comprises a basal thick flow of trachyandesite overlain by several thin flows of olivine basalt with ropy or pahoehoe surfaces. Olivine basalt is the only type found to the west of the waterfall upstream from the Third Fault. These basalts have a maximum exposed thickness of 15 m in the horst between KK (loc. 41) and MCK (loc. 40). Numerous thin, tongue-like sheets, 30 cm to 1 m thick, may be piled on one another, as near HWK (loc. 42), and there are many elongate-to-round lava mounds, or tumuli, between 1 and 4.5 m high.

Perfect preservation of pahoehoe crusts on the basalts shows that they spilled out in rapid succession, almost certainly from a vent nearby rather than from a summit crater of one of the large volcanoes. Directional features in the olivine basalts indicate a vent to the south of the gorge, which fits with the lack of similar flows in exposures to the north. Mafic tuff and lapilli tuff overlying the basalt flows at the Second Fault may well have originated from the same nearby vent.

The flows to the north and south of the gorge are trachyandesites mineralogically similar to that beneath the olivine basalts in the gorge. The only directional feature noted on a trachyandesite flow was smooth, parallel grooving with a bearing of N55° W on the bottom of the flow exposed at the waterfall upstream from the Third Fault.

All of the trachyandesite flows were discharged from Olmoti. The grooving fits with an origin from either Olmoti or Ngorongoro, but K/Ar dates and paleomagnetic measurements place the flows within the eruptive span of Olmoti rather than Ngorongoro (Gromme, et al., 1970). Moreover, the trachyandesites are thickest to the north of the gorge, directly opposite Olmoti, and similar lavas are absent on the outer slopes of Ngorongoro, above the youngest welded tuffs.

Alluvial-Plain Deposits

Alluvial-plain deposits constitute the lower part of Bed I to the west of the Fifth Fault, where they crop out discontinuously over an east-west distance of 9 km. They range in thickness from 12.4 to 21 m except where they thin and pinch out over the top of a buried inselberg (loc. 64). These deposits are chiefly tuff (56%) and claystone (28%) but include limestone, conglomerate, and sandstone. Tuffs form most of the lower two-thirds of the facies, and nearly all of the claystones are in the upper one-third.

Tuffs are characteristically massive or crudely bedded, clayey, and yellowish-brown or yellowish-gray. Rootmarkings are abundant, and most of the tuffs appear to have been homogenized by pedogenic processes. Tuffs are dominantly crystal-vitric, and much of the glass is weathered to montmorillonite clay. The crystal content indicates a quartz trachyte composition. Most of these trachytic tuffs contain a small percent of detritus eroded from the metamorphic basement and the Naabi Ignimbrite. Claystone pellets are relatively common, and most samples have a substantial proportion of sand-size grains, especially pumice, that are coated with claystone giving them a pelletoid texture. An atypical tuff is widespread at and near the base of the alluvial-plain deposits. It is dark gray, generally thinbedded, and consists of shards of basaltic glass.

Claystones form massive beds that are yellowish-gray, pale olive, or yellowish-brown. Most of these have finely textured rootmarkings and contain 5-25% of detrital sand. Sandstones and conglomerates comprise varying proportions of basement and Naabi Ignimbrite detritus. A few of the conglomerates fill narrow, steep-walled channels .5 to 1 m deep. The two channels measured are oriented N70° E and N75° E.

Limestones form massive beds as much as a meter thick that commonly have smooth, sharp upper surfaces and uneven, often gradational lower surfaces. They increase in proportion westward, and in locality 66b they constitute the lower 5 to 8 m of the alluvial-plain deposits. The limestone is characteristically sandy, and most samples also have claystone pellets and clay coatings around sand grains or vitroclasts. These limestones are a type of surface limestone or calcrete that was contaminated with detrital sediment, in part eolian, as it accumulated.

The only faunal remains known from these deposits are suid and elephant teeth, noted in claystone near the top of the facies in locality 71.

The clayey tuffs in the lower part of the facies probably accumulated in a steppe environment where the vegetation was at least seasonally insufficient to prevent considerable reworking of sediment by wind. The degree of weathering and clay formation suggests a rather slow rate of accumulation. Rootmarkings are similar to those produced by grass, and claystone pellets and clay-coated grains indicate eolian transport. The limestones were formed at the land surface and they too received eolian sediment. Conglomerates and sandstones were deposited by streams flowing northeast. The rootmarked sandy claystones probably accumulated on flood plains adjacent to the streams. It should be noted that the alluvial-plain deposits grade upward into lake and lake-margin deposits.

Ngorongoro is the most likely source of the trachytic tuffs in view of their mineralogic similarity to the latest welded tuffs of Ngorongoro. No conclusion has been drawn as to the source of the basaltic vitric tuff.

Environmental Synthesis and Geologic History

The overall paleogeography of Bed I is clear, although several questions remain unanswered. Bed I was deposited in a lake basin at the western foot of the volcanic highlands. The lake either did not have an outlet, or it overflowed infrequently, hence fluctuating in level, alternately flooding and exposing a broad marginal zone (Fig. 6). The

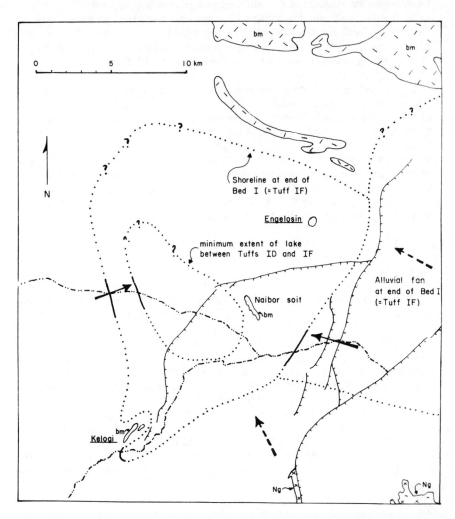

Figure 6.   Paleogeographic diagram showing the outline of the
lake and western extent of the alluvial fan at the
time Tuff IF was deposited.   Also shown is the inferred minimum extent of the lake between the
deposition of Tuffs ID and IF. Solid arrows represent flow directions of streams as based on
channel measurements;   dashed arrows indicate flow
directions as inferred on other grounds.   Lithologic symbols are the same as in Figure 1.

basin was filled by pyroclastic and detrital sediments and by
lava flows. The pyroclastic deposits were erupted from
Olmoti, and the lavas are both from Olmoti and from a vent to
the south of the gorge.

The basin is at least partly tectonic in origin. If the
Naabi Ignimbrite was erupted from Ngorongoro, as concluded
here, then faulting or warping must have reversed the slope
of the land surface from westward to eastward prior to the
deposition of Bed I. The initial paleogeography of the lake
basin remains obscure because of the very limited exposure
below the lavas in the eastern part of the gorge. There is
presently no way to determine whether the welded tuff beneath
the lavas near the Third Fault interfingers with Bed I or
underlies it to the west. However, the 2.4 m of lake-margin
deposits above this ignimbrite almost certainly interfinger
westward with lake deposits. The lavas must have displaced
the eastern shoreline and presumably the overall position of
the lake a substantial distance to the west, where it
remained during the deposition of Bed I. From stratigraphic
and topographic relationships (Fig. 3), it seems likely that
the lavas were erupted at about the time that Tuff IA was
deposited.

Warping or faulting continued during the deposition of
Bed I, as indicated by anomalous paleogeographic relationships at the level of Tuffs IB and IC compared to Tuff IF.
Tuff IF must have been very nearly horizontal when deposited
in the eastern lake-margin terrain, but a reconstructed cross
section of Bed I drawn on this basis places lacustrine clays
between Tuffs IB and IC at FLK topographically higher than
stratigraphically equivalent stream-laid sediment to the east
(Fig. 5). Moreover, if any reasonable water depth is taken
for deposition of Tuff IF in the center of the lake, then
Tuff IB, here deposited in lake water, lies either at the
same or at a higher elevation than it does at the eastern
margin of the lake, where it is a subaerial ash-fall deposit
(Fig. 3). The basin appears to have been tilted slightly
downward to the east, and the area occupied by the lavas may
in addition have been warped into a shallow syncline (Figs.
3, 5).

The average diameter of the lake fluctuated between
limits of about 7 and 25 km during all but the most extreme
periods of desiccation. The lake was widest prior to the ash
flow of Tuff IB, which spread eastward and narrowed the area
subject to flooding. It appears to have been smallest, on
average, between deposition of Tuffs ID and IF, when its
east-west diameter was probably between 5 and 8 km for most

of the time (Fig. 6).   It is reconstructed here as an
elongate body of water oriented northwest-southeast rather
than as a larger, subequant body of water extending much
farther northeast, as inferred previously (Hay, 1970, 1971).
This changed interpretation is based on clasts of phonolite
from Engelosin below Tuff IB at DK and MK which had not been
recognized earlier.   These could not have been transported
for 8 km southward, either by streams or sheetwash, if a part
of the lake intervened between Engelosin and localities in
the gorge with phonolite clasts from it.   Although clasts of
Engelosin phonolite have not been found at higher levels of
Bed I, no basis exists for changing the location or shape of
the lowest part of the basin from the time Tuff IB was deposited to a time between the deposition of Tuffs ID and IF.
The lake expanded rather abruptly at the time Tuff IF was
deposited, for which the shoreline is known with greater
accuracy than any earlier period in the deposition of Bed I
(Fig. 6).

The perennial part of the lake was comparatively shallow
at times of low level as indicated by evidence of wave or
current action at several horizons, and by desiccation cracks
at one level in the central part of the basin.   The wide
marginal zone of intermittent flooding and drying also fits
with a shallow lake.   Water-level fluctuations on the order
of 1.5 to 3.4 m can account for most or all of the features
noted in the lake-margin deposits.   The most significant
measurement in this regard is a stream channel 3.4 m deep
near the western margin of the lake between Tuffs ID and IF.

The perennial part of the lake was saline, alkaline, and
rich in dissolved sodium carbonate-bicarbonate.   Water
flooding the lake-margin terrain varied widely in salinity,
with freshwater near stream inlets along the eastern margin
of the lake.   Frequent fluctuations in level and high
salinity of the lake suggest that its level was controlled by
the balance of inflow and evaporation.   Whether or not the
lake ever overflowed is not known.   The only possible direction of overflow is to the northwest, toward Lake Victoria.

The climate was probably semiarid though wetter than
that in the Olduvai region today.   High salinity and fluctuations in lake level are the best evidence that the climate
was relatively dry over a long period.

Gypsum rosettes ("desert roses") and rare dolomite in
lake-margin deposits are suggestive of dry or saline conditions, at least temporarily.   Alluvial-plain deposits beneath
the lake beds contain much eolian sediment and a few pedogen-

ic calcretes, also suggestive of dry conditions. The fauna includes water- and swamp-dwelling elements as well as plains animals, but these are compatible with a semiarid climate. Urocyclid slugs, found below the level of Tuff ID, appear to be the best evidence clearly indicating a climate appreciably moister than that of today in the same area. Both the fauna and sediments beneath Tuff IF seem to show that the climate was appreciably drier than that prevailing prior to Tuff IC.

The geologic history of Bed I can be briefly summarized as follows:

1. The oldest deposits of Bed I are tuffs overlying the Naabi Ignimbrite in the western part of the gorge. They accumulated on a land surface sloping northeast, and presumably they interfinger with lacustrine sediments in an area not now exposed. Most of the tuffs were probably erupted from Ngorongoro, and the oldest are very likely 2.0 to 2.1 m.y. old.

2. Approximately 1.85 m.y. ago, a large volume of lavas flooded the eastern part of the basin, displacing the lake westward to a position which it occupied for the remainder of Bed I. The bulk of the lavas are trachyandesites from Olmoti; the remainder are olivine basalts erupted from a vent not now exposed which lay to the south of the gorge.

3. The lake was shallow and fluctuated in depth and areal extent, largely in response to changes in the balance of inflow and evaporation. Lacustrine clays accumulated to a thickness of about 26 m in the center of the basin and were deposited widely to the east over the top of the lavas up the deposition of Tuff IB.

4. Tuff IB, erupted 1.79 $\pm$ .03 m.y.a., was the first of a series of deposits produced in a major explosive phase of Olmoti which continued into the lower part of Bed II. These deposits were reworked to form an alluvial fan which displaced the margin of the lake 3.4 km westward from its easternmost limit above the lavas. The fan deposits may well span a period of 20,000 to 50,000 yr, as based on the several paleosols and thickness of lacustrine clays between Tuffs IB and IF in the center of the basin.

5. The climate was probably semiarid but somewhat moister than that in the same area today. A period of relative desiccation seems to be indicated for at least the upper part of the interval between Tuffs ID and IF, roughly 1.75 m.y.a.

Paleogeography of Hominid Activities

    Evidence of hominid activities in Bed I is restricted to
the eastern lake-margin deposits, which accumulated on relatively flat terrain intermittently flooded by the lake (see
Depositional environment, p. 42).   This area supported a
great variety of animal life and a large amount of vegetation.   Lack of game, freshwater, and appropriate vegetation
can account for the apparent absence of evidence for hominid
activities in other time-equivalent lithologic facies of Bed
I.   Within the eastern lake-margin deposits, the present
distribution of hominid materials differs considerably between upper and lower units (Fig. 5).   In the lower unit,
hominid remains and archaeologic materials are confined to
the north side of the gorge between localities 10 and 14.
These sites lay in the eastern or inland half of the terrain
intermittently flooded by the lake.   Most of these come from
the upper most 2 m of sediment beneath Tuff IB between localities 10 and 14.   The excavation at DK (loc. 13) provided
some detailed information of paleo-geographic significance
(M.D. Leakey, 1971).   The stone cir-cle and occupation site
overlie a paleosol developed partly on the eroded surface of
a tuff and partly on the basalt where it rose above the level
of the tuff.   The surface of the tuff showed a number of
narrow, steep-sided channels 45 to 60 cm deep that resemble
game trails leading to the edge of the lake in Ngorongoro
(ibid., p. 12-23).   Fossil Papyrus(?) rhizomes and large
quantities of crocodile remains testify to permanent water
nearby.

    Above the level of Tuff IB, hominid remains and arti-
facts are found only to the west, in the vicinity of FLK and
HWK.   These sites appear to have positions paleogeographi-
cally more or less comparable to the site at DK.   The occupation site at FLK-NN (site 38b, level 3) is on the weathered
surface of a thin claystone above Tuff IB.   "The many rootlet
holes and reed casts in Tuff IB and occurrence of numerous
fish and amphibian remains, together with bones of waterfowl,
indicate that the site was situated near the shores of a lake
or by a swamp" (ibid., p. 42).   The site at FLK, termed the
"Zinjanthropus floor," lies on a lacustrine claystone beneath
Tuff IC.   The top of the claystone is a weakly developed
paleosol, slightly uneven, and cut by a small channel of
undetermined origin.   Tuff IC contains abundant rootcasts of
marshland vegetation.   At FLK-N, the topmost 1.5 m of claystone beneath Tuff IF has yielded five implement-bearing
levels.   The lowest level (no. 6) was a butchering site where
artifacts were associated with the skeleton of an elephant.
The highest level (nos. 1 and 2) was an occupation site comparable in many respects to the "Zinjanthropus floor."   Artifact concentrations at the intermediate levels are probably

reworked. These claystones were exposed at the surface too briefly to develop paleosols. The shoreline may well have been a kilometer or so to the west of these sites for much of the time represented these levels at FLK-N (Fig. 5). Scattered artifacts have been encountered in both excavations and natural exposures at other levels between Tuff IC and level 6 at FLK-N. They seem to be most widespread in the massive tuff that overlies Tuff ID and which contains diatoms, ostracods, and rootmarkings of marshland vegetation (sites 41e, f).

Acknowledgements

Field work at Olduvai Gorge has been supported by grants from the National Geographic Society and the National Science Foundation (G-22094); laboratory work and writing were aided by research professorships in the Miller Institute for Basic Research (Berkeley). Dr. L.S.B. Leakey provided the opportunity to study the Olduvai deposits and helped me to obtain financial support for field work. I am grateful to Dr. M.D. Leakey for use of the camp facilities at Olduvai and for generous assistance in various aspects of the geologic work between 1962 and the present. Discussions with Dr. Glynn L. Isaac have been a fruitful source of information and interpretation. Robert N. Jack is much to be thanked for diffractometer analyses, and I am indebted to S.J. Chebul for the thin sections used in microscopic study. Unpublished isotope analyses were generously made available by Dr. J.R. O'Neil of the U.S. Geological Survey.

References

Brock, A., Hay, R.L., and Brown, F.H., 1972, Magnetic stratigraphy of Olduvai Gorge and Ngorongoro, Tanzania (abstract): Geological Society of America Abstracts with Programs for 1972, p. 457.

Curtis, G.H., and Hay, R.L., 1972, Further geologic studies and K-Ar dating of Olduvai Gorge and Ngorongoro Crater, in Bishop, W.W., and Miller, J.A., eds., Calibration of Human Evolution, Scottish Academic Press, Edinburgh and London, p. 289-301.

Fleischer, R.L., Price, P.B., Walker, R.M., and Leakey, L.S.B., 1965, Fission-track dating of Bed I, Olduvai Gorge: Science 148, 72-74.

Graf, D.L., 1969, Geochemistry of carbonate sediments and sedimentary carbonate rocks, Part IV-A, Isotopic composition and chemical analyses: Illinois Geological Survey, Circular 308, p. 42.

Greenwood, P.H., and Todd, E.J., 1970, Fish remains from Olduvai, in Leakey, L.S.B. and Savage, R.J.G., eds., "Fossil Vertebrates in Africa", Vol. 2, Academic Press, London and New York, p. 225-241.

Gromme, C.S., and Hay, R.L., 1971, Geomagnetic polarity epochs: age and duration of the Olduvai normal polarity epoch: Earth and Planetary Science Letters 10, p.179-185.

Gromme, C.S., Reilly, T.A., Mussett, A.E., and Hay, R.L., 1970, Paleomagnetism and potassium-argon ages of volcanic rocks of Ngorongoro Caldera, Tanzania: Geophysical Journal of the Royal Astronomical Society 22, p. 101-115.

Hay, R.L., 1966, Zeolites and zeolitic reactions in sedimentary rocks: Geological Society of America Special Paper 85, p. 130.

Hay, R.L., 1968, Chert and its sodium-silicate precursors in sodium-carbonate lakes of East Africa: Contribution to Mineralogy and Petrology 17, p. 255-274.

Hay, R.L., 1970, Silicate reactions in three lithofacies of a semiarid basin, Olduvai Gorge, Tanzania, in Morgan, B.A., ed., Mineralogical Society of America Special Paper No. 3, p. 237-255.

Hay, R.L., 1971, Geologic background of Beds I and II. "Olduvai Gorge": Vol. 3, Leakey, M.D., Cambridge University Press, p. 9-18.

Isaac, G.L., 1967, The stratigraphy of the Peninj Group-early middle Pleistocene formations west of Lake Natron, Tanzania, in Bishop, W.W. and Clark, J.D., eds., Background to Evolution in Africa", University of Chicago Press, p. 229-257.

Leakey, L.S.B., 1965, Olduvai Gorge, 1951-61, Cambridge University Press, p. 118.

Leakey, M.D., 1971, Olduvai Gorge, Vol. 3., Cambridge University Press, p. 306.

O'Neil, J.R., and Hay, R.L., 1973, O^{18}/O^{16} ratios in cherts associated with the saline lake deposits of East Africa: Earth and Planetary Science Letters 19, p. 257-266.

Price, W.A., 1963, Physiochemical and environmental factors in clay dune genesis: Journal of Sedimentary Petrology 33, p. 766-778.

Reck, H., 1951, A preliminary survey of the tectonics and stratigraphy of Olduvai, in Leakey, L.S.B., ed., Olduvai Gorge, Cambridge University Press, p. 5-19.

Selley, R.C., 1970, Ancient Sedimentary Environments, Cornell University Press, p. 224.

Verdcourt, B., 1963, The Miocene nonmarine mollusca of Rusinga Island, Lake Victoria, and other localities in Kenya: Palaeontographica Bd. 121, Part A, p. 1-37.

Howard J. White, Daniel R. Burggraf, Jr.,
Russell B. Bainbridge, Jr., Carl F. Vondra

3. Hominid Habitats in the Rift Valley: Part 1

Abstract

Lake Turkana, in the northwestern quarter of Kenya, is one of many African lakes developed in depressions formed by the crustal downwarping associated with Cenozoic continental rifting. The lake occupies a half-graben bounded on the west by a zone of major normal faulting and on the east by a westward dipping monoclinal flexure. During the Pliocene and Pleistocene epochs, the eastern shoreline of the paleo-Lake Turkana fluctuated in response to movement along the marginal faults which raised or lowered the half-graben floor. Consequently, more than 300 m of lacustrine, transitional and fluvial sediments were deposited as a westwardly prograding sequence which recorded the gradual shift of the lake's eastern shoreline to the west. This transition is recorded by four major lithofacies which are: (1) the laminated siltstone facies, (2) the arenaceous bioclastic carbonate facies, (3) the lenticular, fine-grained sandstone and lenticular-bedded siltstone facies, and (4) the lenticular conglomerate, sandstone, and mudstone facies. Based on lithologies, bed geometries, primary sedimentary structures, and fossil content, corresponding depositional environments for each of the lithofacies are: (1) prodelta and shallow-shelf lacustrine, (2) littoral lacustrine--beach, barrier beach, and associated barrier lagoons, (3) delta plain--distributary channel, levee, and interdistributary flood basin, and (4) fluvial channel and flood basin.

Introduction

East African rift-valley sediments have yielded important paleo-anthropological evidence critical for the documentation of hominid evolution. The discoveries from the sediments east of Lake Turkana, Kenya, have stimulated an active research interest. The understanding of the complex fluvio-lacustrine stratigraphy in which these paleontological and archaeological sites are found is the objective of both regional and local geologic investigations.

This first of two parts on the rift-valley setting presents an overview of the geology of the East Turkana Basin and, in particular, the stratigraphy of the Koobi Fora Formation. Part II focuses on the lithofacies and paleoenvironmental interpretations of the Karari Escarpment's archaeological and anthropological sites. This symposium of the American Association for the Advancement of Science provides a forum for detailing and comparing the East African rift setting with the geologic settings of world-wide occurrences of prehistoric hominid habitats.

Previous Work

Literature about Lake Turkana (formerly Rudolf) and its surrounding region dates prior to the turn of the century. Hohnel et al. (1891) reported on the first European journey to the lake by the Austrian Count Teleki in 1888. Observations by Hohnel encouraged E. Suess (1891) to hypothesize tectonic continuity between the Red Sea fractures and the inland East African graben system.

These earliest expeditions (Hohnel, 1894; Smith, 1896; Harrison, 1901) concentrated on traversing previously unexplored territory and reported only briefly on the geological and paleontological significance of the rift-valley sediments. Later the Cambridge Expedition of 1930 (Fuchs, 1934) and the Lake Rudolf Rift Valley Expedition of 1934 (Fuchs, 1935, 1939) added further confirmation of the significance.

Camille Arambourg led the first paleontological-collecting expedition into the region in 1932 (Arambourg, 1935). Since that time, most of what is known of the geology of the upper Cenozoic deposits has been accumulated in conjunction with paleontological expeditions to various margins of the lake. From 1963 to 1968, Harvard University expeditions completed systematic areal reconnaissance of the Miocene Turkana Grits along the southwest margin (Patterson et al., 1970; Behrensmeyer, 1974). Princeton University conducted similar studies of the Turkana Grits north of the Harvard area. The success of Arambourg in his early expeditions to the Omo River gave impetus for the 1967 International Paleontological Research Expedition. The stratigraphy and depositional environments of the Plio-Pleistocene Omo sediments were studied by several members of the expedition

(Butzer and Thurber, 1969; DeHeinzelin et al., 1971; Brown, this volume). The Geological Survey of Kenya has also published detailed geologic maps of the western and southern margins of the Lake Turkana area (Dodson, 1963; Joubert, 1966; Walsh and Dodson, 1969).

The present review stems from the 1968 East Rudolf Research Expedition organized by R.E.F. Leakey, Director of the National Museums of Kenya. A.K. Behrensmeyer initiated the geologic study of the East Turkana area in conjunction with this expedition and as a result introduced an informal stratigraphic framework in 1969 (Leakey et al., 1970). Geologists from Iowa State University joined the expedition and began a systematic study and regional mapping program in 1970 (Vondra et al., 1971). A formal stratigraphic nomenclature proposed by Bowen and Vondra (1973) and regional geologic maps (1:24,000) were completed by Bowen (1974) with the assistance of others (Acuff, 1976). Further descriptions of the stratigraphy and depositional environments include Vondra and Bowen (1976, 1978), Acuff (1976), Bainbridge (1976), Burggraf (1976), Vondra and Burggraf (1978), Findlater (1976, 1978), Frank (1976), and White (1976). Mathisen (1977) delineated heavy mineral associations within the basin.

An absolute dating system based on radioisotopic analyses of selected minerals of volcanogenic origin was developed by Fitch et al. (1974). Major tuff levels were dated using conventional K-Ar (Fitch et al., 1976; Fitch and Miller, 1976; Curtis et al., 1975), $^{40}Ar/^{39}Ar$ age spectrum techniques (Fitch and Miller, 1976; Fitch et al., 1976), and fission-track analyses (Hurford et al., 1976). Additionally, paleomagnetic (Brock and Isaac, 1974; Hillhouse et al., 1977) and paleontologic (Harris, 1976; Harris and White, 1977) studies have led to some controversy relating biostratigraphic, time-stratigraphic, and lithostratigraphic frameworks between separate areas within the basin and between the East Turkana Basin and the Omo Basin farther north (see in particular Harris and White, 1977; Hillhouse et al., 1977).

Since 1968, the late Upper Cenozoic sediments surrounding Lake Turkana have been the setting for one of the most comprehensive, interdisciplinary paleoanthropological research efforts ever undertaken. The reports of the numerous expeditions have been assimilated into the proceedings of a symposium entitled "Earliest Man and Environments in the Lake Rudolf Basin - Stratigraphy, Paleoecology, and Evolution" (Coppens, Howell, Isaac, and Leakey, 1976). Three additional compilations incorporating stratigraphic research at East Turkana are Background to Early Man in

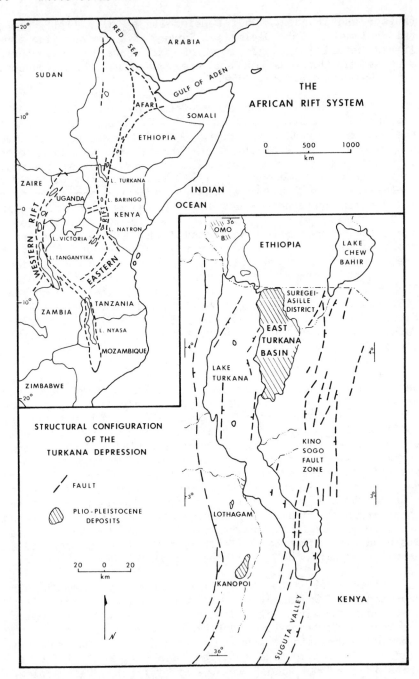

Figure 1.   The East African Rift System showing location of major rift-valley lakes and the principal elements of the Turkana Depression.

Africa, W.W. Bishop, ed., (Vondra and Bowen, 1978), <u>Fluvial</u>
<u>Sedimentology</u>, A. D. Miall, ed., (Vondra and Burggraf, 1978),
and <u>The Koobi Fora Research Project</u>, Vol. 1, M.G. Leakey and
R.E.F. Leakey, eds., (Findlater, 1978).

Geologic Evolution of the East Turkana Basin

The geologic setting and tectonic framework of the East
African Rift System have received considerable attention due
to the interest generated by the concept of plate tectonics
(Baker, 1970; McKenzie <u>et</u> <u>al</u>., 1970; Baker and Wohlenberg,
1971; Baker <u>et</u> <u>al</u>., 1972; McConnell, 1972). The configuration of the Eastern (Gregory) Rift System in Kenya and the
Western Rift System in Tanzania and Uganda is well known
(Fig. 1). The two systems join to the south in Mozambique
and again to the north through Ethiopia and the Afar junction
with the Red Sea and Gulf of Aden.

Linear and sometime diffuse topographic depressions are
indicative of these initial plate-rifting movements and are
occasionally occupied by shallow lakes, e.g., lakes Turkana,
Baringo, Natron, and Eyasi. Hominid artifacts of these are
encased by Upper Cenozoic sediments which surround these
lakes. These document favorable ecosystems for habitat and,
with the prolific anthropological and paleontological finds,
provide added urgency for detailed basinal analyses of the
intra-rift stratigraphy.

Precambrian metasediments comprise the basement complex
exposed west of the lake along the Uganda Escarpment and to
the north and northeast. Walsh and Dodson (1969) described
these rocks as leucocratic migmatitic gneisses to patchy
mesocratic migmatites. The metasediments are considered part
of the north-south Mozambique Belt and the resultant of
several older Precambrian orogenic cycles. The rock record
of the region surrounding Lake Turkana excludes the interval
of time spanning the events associated with the folding and
metamorphism of the basement system and prior to rift-related
deposition of the Tertiary Turkana Grits (Walsh and Dodson,
1969). Preserved on the western side of the Rift Valley,
these massive quartzites are believed to be earliest Miocene
in age.

Initial rifting apparently began with uplift in the
Ethiopian-Arabian region during the late Eocene (Baker <u>et</u>
<u>al</u>., 1972). Early Miocene domical upwarping propagated
rifting southward into both Ethiopia and Kenya. The
resulting triangular Turkana Depression between the two domal
uplifts was partially filled with Miocene (17.2 $^+$ 1.8 m.y.
B.P.) to Pliocene ignimbrites, basalt lavas, and minor sedi-

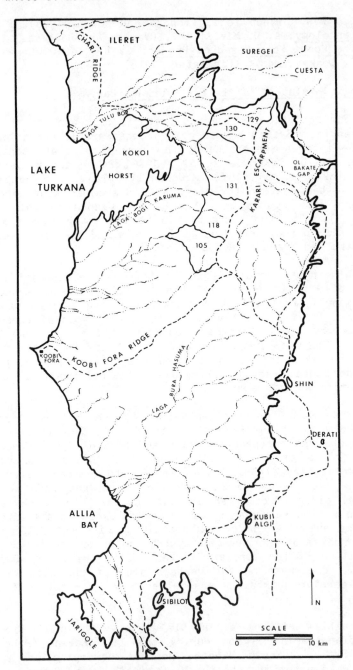

Figure 2. Physiographic map of the East Turkana Basin and
surrounding volcanic highlands. Paleontologic-
collecting localities for the Karari Escarpment
referred to in the text are also shown.

mentary intercalations (Fitch and Miller, 1976; Watkins; personal communication). Late Miocene and Pliocene faulting defined the principal elements of the eastern rift in northern Kenya, namely the Suguta Valley and Lake Chew Behir grabens shown in Figure 1. Intermediate half-graben faulting created the diffuse northeast trending faults of the Kino Sogo fault zone (Baker et al., 1972). Cerling and Powers (1977) interpret the Kino Sogo fault zone to have been formed subsequent to the Turkana Depression from lateral south-westerly migration of crustal deformation.

Beginning in the late Miocene, deformation created north to north-northeast trending faults antithetic to the Kino Sogo fault zone. It is within this meridional half-graben system that deposition of the Pliocene and Pleistocene East Turkana sediments occurred in association with marginal Pliocene flood basalts (3.8 $^+$ 0.4 m.y. B.P. to 2.20 $^+$ 0.44 m.y. B.P.) (Fitch and Miller, 1976). Evidence of faulting in the early Pliocene and Pleistocene has been cited by Walsh and Dodson (1969). The continuance of tectonic activity into the late Pleistocene and Holocene is further suggested by the occurrence of Galana Boi Beds on the faulted Kokoi Horst 120 m above equivalent discontinuous occurrences elsewhere in the basin (Bowen, 1974).

Miocene and Pliocene volcanic associations define the northeastern, eastern, and southern limits of the East Turkana Basin. The Suregei Cuesta (Fig. 2) is a westward-tilted monoclinal flexure of volcanics rimming the basin on the northeast and east and dipping beneath the basin sediments to the west. South of the cuesta, the prominent land-forms of Shin, Derati, and Kubi Algi are Miocene volcanic masses surrounded by younger Pliocene lava flows (Fitch and Miller, 1976); Sibilot and Jarigole are also volcanic in origin. Within the basin, distinctive landmarks include the Kokoi Horst, a faulted complex of Upper Pliocene volcanic rocks and lacustrine sediment trending northeastward, north of the Koobi Fora Ridge. The ridge itself and the Karari Escarpment are both cuestas of interbedded and capping-resistant sedimentary strata. Maximum relief to the top of the Suregei Cuesta is approximately 475 m above mean lake level at 375 m altitude. Presently, all drainage of the East Turkana embayment flows westward into the lake. Besides the major physiographic features of East Turkana, Figure 2 also shows the boundaries of arbitrarily designated paleontologic collecting areas established to facilitate on-going research efforts and documentation of specimen finds for future reference.

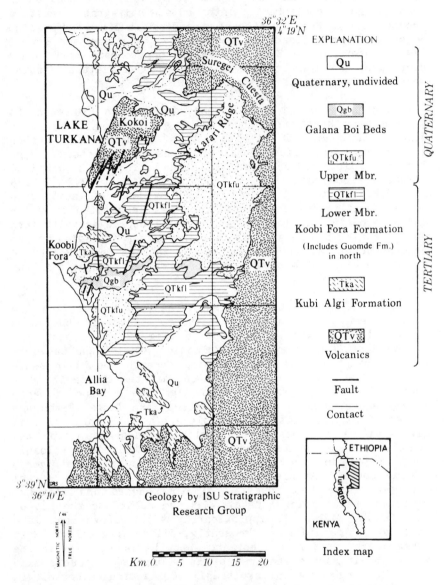

Figure 3. Geologic map of East Turkana (reprinted by permission from Vondra, Burggraf, and White (1978), Fig. 4).

Basinal Stratigraphy

The Upper Cenozoic strata of the East Turkana Basin were formally established by Bowen and Vondra (1973). The section constitutes a 325-m fluvio-lacustrine complex which rests unconformably on Miocene and Pliocene volcanics. This was subdivided into three units of formational rank named the Pliocene Kubi Algi Formation, the Plio-Pleistocene Koobi Fora Formation, and the Pleistocene Guomde Formation. Discontinuous Holocene lake deposits were named the Galana Boi Beds. Subsequent publications have provided more complete descriptions of the strata and lithologic facies, as well as interpretations of their environments of deposition (Bowen, 1974; Bowen and Vondra, 1974; Vondra and Bowen, 1976, 1978; Vondra and Burggraf, 1978; Vondra, Burggraf, and White, 1978; Findlater, 1976, 1978). Figure 3 is a modified reproduction of the original geologic maps of Bowen (1974).

The outcrop pattern of East Turkana is the result of late Quaternary tectonism and concomitant erosion. Exposures are generally restricted to ridges capped by resistant tuffs, carbonate-cemented sands and carbonates, and to interfluves of ephemeral streams. A thin veneer of Recent stream alluvium and patches of Holocene lacustrine Galana Boi Beds cover large areas of the sedimentary basin precluding establishment of indisputable basin-wide physical correlation of all exposures. However, sufficient field studies have been accomplished throughout East Turkana (Bowen, 1974; Findlater, 1977) for secure correlations of the stratigraphic section. These basinal correlations are generalized in Figure 4, illustrating the thicknesses measured in the Ileret, Karari, Koobi Fora, and Alia Bay areas but only abbreviating the complex stratigraphy. Lithostratigraphic sectioning and correlation are especially detailed along the Karari Escarpment within which the principal archaeological industries are found (Bainbridge, 1976; Burggraf, 1976; Frank, 1976; White, 1976). The Karari Escarpment will be the principal focus for this chapter.

Kubi Algi Formation

The oldest rift sediments at East Turkana are most extensive in discontinuous exposures south of the Laga Bura Hasuma (Laga = river) in the Allia Bay area. The section is named after Kubi Algi, a Miocene rhyolitic plug to the east along the basin margin. North of Allia Bay scattered outcrops occur between the Koobi Fora Ridge and the Kokoi Horst. A maximum of 154 m has been measured near Jarigole on the extreme southern margin (Bowen, 1974; Acuff, 1976). Thickness decreases from basin edge to center primarily due to

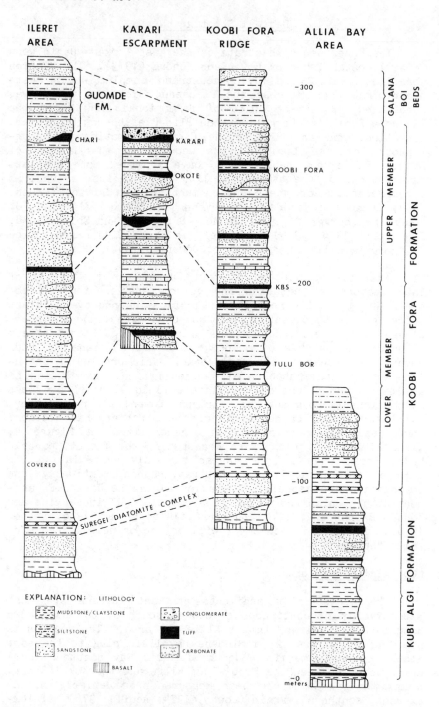

Figure 4. Generalized basinal stratigraphy. Modified
from Bowen and Vondra (1973).

alluvial fan development proximal to surrounding volcanic highlands.

Bowen (1974) and Acuff (1976) have investigated the type locality centered 13 k southwest of Kubi Algi. Here the 98-m section has been described as a stacking of ten conglomeratic to claystone, fining-upward cycles (Acuff, 1976), individual cycles varying in thickness. A typical cycle begins with a basalt clast conglomerate, continues fining with a trough to planar cross-bedded feldspathic litharenite in which the scale of cross-bedding decreases vertically. This is followed by lenticular-bedded siltstone displaying calcareous root casts and concretions, and finally capped by claystones (Acuff, 1976). The prominent sand intervals may be as much as 3.5 m thick (Bowen, 1974) and contain abundant invertebrates, e.g., ostracods, gastropods and pelecypods, and occasional fish and mammal remains. Cerling (1979) has recognized freshwater diatomite beds in the upper portion of the Kubi Algi. In the marginal Jarigole section, fanglomerates with basal erosional contacts reflect an influx of local acidic volcanic detritus with abundant ignimbrite gravels. This alluvial fan wedge comprises fewer cycles but each of greater thickness than those basinward.

Several volcanic tuff intervals occur in the Kubi Algi. Two in particular, the Hasuma and Allia Tuffs, have been used as prominent marker horizons for local correlation in the Allia Bay area. Dating by potassium-argon methods, Fitch and Miller (1976) calculated a Pliocene age range of 3.8 to 4.9 million years (m.y. B.P.) for the Kubi Algi; this is consistent with collected vertebrate fossil assemblages (Maglio, 1972). The Kubi Algi Formation disconformably overlies or is in fault contact with Miocene and Pliocene volcanics.

Koobi Fora Formation

The extensive exposures of the Koobi Fora Formation have yielded more than 100 fossil hominids, a wealth of other vertebrate specimens, and the world's oldest archaeological sites yet discovered. As such, the formation has received the greatest interest from the interdisciplinary Koobi Fora Research Project. Thickness varies considerably basin-wide, the greatest being 210 m in the Ileret area. Eastward from the Koobi Fora spit, for which it was named, the formation decreases from 175 m to 135 m along the Karari Escarpment and to 47 m near Derati.

The Koobi Fora, in its type section, consists of small-scale cross-bedded sublitharenites and laminated siltstones interbedded with minor pebble conglomerate, bioclastic and biolithic carbonate, and tuff (Bowen, 1974). Two members, designated simply as Lower and Upper, are recognized (Bowen and Vondra, 1973; Vondra and Bowen, 1978). Previously, Koobi Fora strata in the Ileret area had been termed Lower and Ileret Members because of difficulty in physical correlation with the Lower and Upper Members recognized elsewhere (Bowen and Vondra, 1973; Bowen, 1974; Vondra and Bowen, 1976). Accordingly, the stratigraphic boundary between the members had been redefined for intrabasinal consistency (Vondra and Bowen, 1978). Part I of the rift-valley setting will present in some detail the stratigraphy of the Lower Member particularly as it exists along the Karari Escarpment. Part II will detail the lithostratigraphy of the Upper Member associated with the archaeological sites of the Karari Industries.

An apparently conformable relationship exists between the Kubi Algi and Koobi Fora Formations. The base of the Suregei diatomite complex, where present, originally identified as tuffaceous, defines the contact. Where the Kubi Algi is absent, the Koobi Fora Formation rests disconformably on Miocene and Pliocene volcanics.

Lower Member. Strata of the Lower Member outcrop east of the Chari Ridge in a continuous band from the Karari Escarpment to the Koobi Fora Spit, and over extensive areas bordering the Laga Bura Hasuma (Fig. 3). The member varies in thickness from 123 m just east of Ileret to 78 m in the Koobi Fora Ridge.

Two prominent tuff marker horizons occur within the Lower Member. These are the Tulu Bor and the KBS. Though discontinuous, the tuffs provide a basis for correlation between distinct areas. These tuffs and those of the Kubi Algi Formation and the Upper Member serve as the primary correlation units within East Turkana. Each tuff may occur as a complex interbedded with thin sedimentary intercalations. Within the limitations of radiometric dating techniques, they are considered basin-wide time-equivalent units that are easily recognizable in the field (Findlater, 1976). As previously stated, the tuffs are discontinuous along their stratigraphic level due to penecontemporaneous depositional conditions or post-depositional erosion. Individual tuff beds display expected variations in thickness, continuity along strike, and internal primary structures.

As a group, the tuffs exhibit a comparatively purer and finer-grained basal zone a few centimeters thick. This is overlain by silt- to coarse-sand-sized glass shards with a varying admixture of volcanic crystal fragments such as sanidine-anorthoclase, pyroxene, ilmenite, and bypyramidal sanidine-anorthoclase, pyroxene, ilmenite, and bypyramidal high quartz plus some zircon and volcanic rock fragments (Findlater, 1976). Colors vary from white and grayish shades to pale green and orange-brown depending on the extent of reworking and terriginous mixing. The primary small-scale cross-bedding to ripple and planar laminations suggest aqueous transport and reworking from primary airfall deposits (Bowen, 1974; Findlater, 1976). Findlater (1976) presents a hypothesis for ash-choked stream transport during and following pyroclastic eruptions. Pumice clasts to 25 m diameter (Findlater, 1976) are present in several of the tuff horizons and have supplied sanidine-anorthoclase crystals for K-Ar radiometric dating. Diagenesis has produced authigenic clay minerals, zeolites, and carbonates locally cementing the units and filling root casts and crotovina present within the upper gradational contact.

Below the Tulu Bor Tuff in the Ileret area, molluscan subarkoses capped by bioclastic carbonates are overlain by a 53-m-thick, planar cross-bedded sublitharenite and mudcracked siltstone complex. Above the tuff to the east, laminated siltstones intercalated with thin lenses of gastropods grade vertically into several repetitions of pebble conglomerate to claystone gradations. These cycles interfinger westward along the Koobi Fora Ridge with ripple-laminated litharenite, bioclastic carbonates, siltstones, and local intraformational conglomerates. Above the Tulu Bor Tuff near the Laga Bura Hasuma, the interbedded siltstones and bioclastic carbonates interfinger with large-scale, planar cross-bedded sands.

Figure 5 diagrams the lateral and vertical stratigraphic relationships of the Lower Member from area 129 to area 105 southward along the Karari Escarpment. Lower Member rocks exposed along the escarpment envelop two prominent marker beds, the Tulu Bor and the KBS tuffs. The characteristics of the sediments between the two markers vary laterally. The sediments at the north end are of a coarser grain and are more poorly stratified than their lateral equivalents in the south. This change in character is a result of increased distance from the basin margin and basinward progradation of migrating depositional environments.

The Koobi Fora Formation rests on deeply weathered vesicular basalts of the backslope of the Suregei Cuesta in northern area 129. The Lower Member here thins to 20 m and is deposited on a surface of some relief developed on the basalt. The sequence consists of the Tulu Bor Tuff, claystone, poorly sorted conglomerate interbedded with poorly stratified, pebbly mudstone, and thick sandstone with laterally equivalent mudrocks.

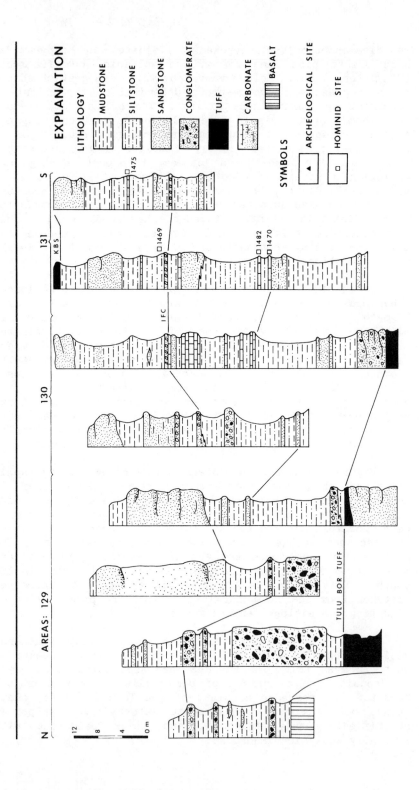

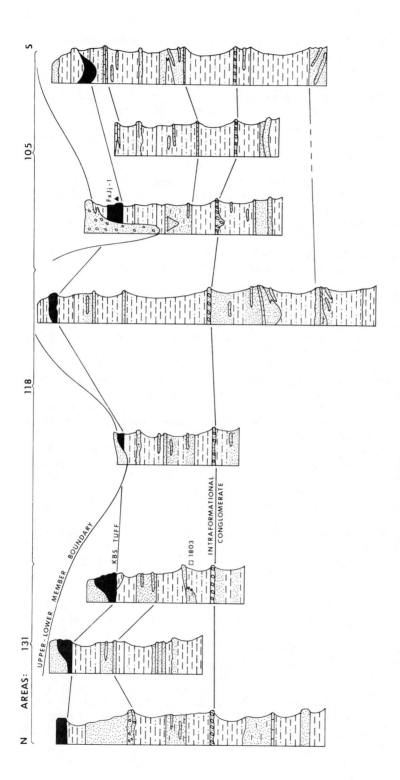

Figure 5. Stratigraphic correlation of the Lower Member, Koobi Fora Formation along the Karari Escarpment.

The Tulu Bor Tuff is fine- to very fine-grained and is well-sorted with only a small admixture of sand or silt. It ranges in color from light grayish green (5 GY 8/1) to very light gray (N8) and in thickness from 3 to 5.5 m. In places, the tuff is deeply weathered and no sedimentary structures are observable. In other places, the tuff consists of two parts with differing sedimentary structures. The basal part contains both small-scale planar and trough cross-bedding as well as climbing ripple-lamination. The upper part is thoroughly bioturbated, with prominent smooth tube trace fossils. Laterally, the tuff interfingers with a fine-grained, moderately well-sorted, yellowish brown (10 YR 6/3), tuffaceous feldspathic litharenite, finely laminated, and also extensively bioturbated. Lenses within the sand are coarse, poorly sorted, and display trough cross-bedding. The tuff was deposited on a surface of relief and overlies both massive poorly indurated sands and reddish-brown mudstones.

Interbedded basalt clast conglomerates and pebbly mudstones lie above the Tulu Bor Tuff in area 129. The sediments have a maximum thickness of 31.5 m. The lower occurrence of the conglomerate is a lenticular channel complex displaying basal erosional contacts with underlying sediments. These complexes are thick (16 m) and are large-scale, planar, and trough cross-bedded. The upper occurrences are finer grained, thinner, and tabular in shape. Each occurrence of the conglomerate displays a series of coarsening-upward cycles. At the base of the lower conglomerate is a gravelly sandstone or imbricated basalt pebble conglomerate which coarsens upward to a cobble and boulder conglomerate. Upper occurrences of the conglomerate consist of sand at the base, coarsening upward to pebble conglomerates. The upper conglomerates are horizontally bedded and contain small channels with small-scale trough cross-bedding within them. Thin layers of lacustrine gastropods are frequently found at the base of the upper conglomerates. Each of the conglomeratic units is very poorly sorted with an interstitial matrix of clay or silt and is poorly indurated. Moderate brown (5 YR 4/4) pebbly mudstones lie between the conglomerates. The mudstones are very limonitic and contain stringers of gypsum and carbonate in addition to the basalt pebbles. Contacts between the mudstones and conglomerates are gradational.

Southward along the escarpment, the oldest rocks exposed lie above the Tulu Bor stratigraphic level and comprise a mudrock sequence. Color gradations from moderate brown (5 YR 3/4) to light gray (10 YR 7/2) correspond respectively with the textural variation. Mudstone layers typically show indistinct parallel and ripple lamination to thin-bedding and irregular to vertically prismatic fracture. Frequently, thin

beds of laminated mudstone are interlayered with similarly
bedded siltstone. Ripple-laminated sandstone lenses also
occur in the siltstone. Basal contacts are generally sharp
and regular, but gradational or slightly irregular boundaries
are occasionally observed. Bedding plane structures include
infrequent but distinct in situ ripple marks. Though not
abundant, gastropod and pelecypod fossils occur in thin local
lenses. Several secondary features commonly present in Lower
Member mudrocks include thin (0.5 to 2 cm) gypsiferous vein
fillings, black, dendritic manganese oxide staining and salt
(gypsum and halite) encrustations on fracture surfaces, limonitic staining along bedding planes, and nodular, goethite-
and calcite-cemented concretions. The concretions are con-
sistently associated with the limonitic banding along well-
defined bedding plane surfaces.

The calcareous feldspathic litharenite grading into arenaceous bioclastic carbonate sequences are locally prominent
within the lower mudrock section in the south (area 118-105).
The sandstones vary in color from very pale orange (10 YR
8/2) to dark yellowish-orange (10 YR 6/6). Sorting varies
with silt content, becoming more poorly sorted with an increasing admixture of fines. Basal contacts of the sequences
are gradational to locally sharp with the underlying siltstones. The sandstones of each sequence exhibit a distinguishing complex of primary sedimentary structures. Large-
scale planar foresets dipping 4° to 9° west-southwest
characterize the sandstones where they are thickest. The
planar foresets consist of smaller large- and small-scale
trough- and ripple-drifted cross-beds. Laterally the sandstones thin and interfinger with siltstone.

Bioclastic carbonates, capping the large planar foresets, thicken from an average of 0.1 m to locally over 1 m.
The carbonates are grayish-orange (10 YR 7/4) weathering to
dusky yellowish-brown (10 YR 2/2) and are composed of a
packed gastropod framework in a matrix of calcite-cemented
sand. Virtually all of the molluscan tests have undergone
extensive recrystallization and dissolution resulting in
arcuate voids surrounding internal mold fillings in the arenaceous matrix. Pelecypods, ostracods, and disarticulated
vertebrate fauna (fish, crocodile, and turtle) are also present. Mutela sp. casts occur in the subjacent sandstones.
The sandstone overlying the bioclastic carbonate is generally
thin, less than 0.2 m, and structureless but distinctive
because of its dark yellowish orange (10 YR 6/6) color.
Contact with this sandstone is sharp but irregular. The
large-scale cross-bedding of the subjacent sandstone continues through the bioclastic carbonate, terminating at the
upper contact. The sequences vary from dominantly loose sand

to very well-indurated bioclastic carbonate depending on the abundance of fossils and degree of calcite induration. Maximum thickness of each sequence is 4 m.

Approximately 20 m below the KBS Tuff is a distinctive stratigraphic interval from area 130 to 105 of intraformational conglomerates consisting of dark yellowish orange (10 YR 6/6) granule- to cobble-sized intraclasts of limonitic siltstone. Locally in area 105 the clasts attain a maximum diameter of 10 to 15 cm, are plate shaped, and frequently encrusted with an algal carbonate veneer up to 2 cm thick. An indistinct imbrication is noted where the large clasts are abundant. Clast size decreases rapidly to the north, granule- to pebble-sized fragments being common. These indurated beds interfinger and disappear to the east in area 130 into sandy, pale yellowish brown (10 YR 6/3) lenticular-bedded siltstone similar to those overlying the interval. The matrix of the conglomerates is arenaceous calcite, locally very fossiliferous with abundant molluscan casts and molds, and occasional disarticulated vertebrate fragments (fish and crocodile). The sharp basal contact exhibits more than 1 m of relief. Worm burrows occasionally occur on the upper surface. The usually high degree of induration and the persistent lateral continuity make the conglomerate extremely valuable for mapping and local correlation (Fig. 5).

Overlying the conglomerate, several distinctive strata and interbedded complexes occur within the dominantly mudrock section. Although texturally and mineralogically similar to the mudrocks occurring in the lower part of the Lower Member, distinct differences are present. Locally above the conglomerate, approximately 3 m of lenticular-bedded mudstone and fine-grained sandstone are exposed. The sandstone lenses, less than 5 cm thick, display typical small-scale cross-bedding and ripple lamination. Laterally as well as vertically, the beds become repetitive, 0.2 m thick fining-upward sequences. Each sequence begins as thin-bedded to ripple-laminated pale yellowish brown (10 YR 6/2) mudstone and grades into thin (2-4 cm) beds of faintly parallel-laminated moderate brown (5 YR 3/4) claystone. Basal contacts for each sequence are generally sharp to slightly gradational. Thickness of the rhythmic mudstone locally exceeds 8 m.

Overlying the rhythmic mudstone is an interfingering complex of locally large-scale cross-bedded sandstone and bioclastic carbonate and an extensive sequence of coarser grained rhythmic sandstone and mudstone. Equivalent stratigraphic levels to the north (area 130-131) contain thick lenticular sandstones and laterally equivalent claystone.

The sands are pale yellowish brown (10 YR 6/2) lithic arkoses medium-grained, large-scale trough cross-bedded at their base grading upwards to fine-grained, ripple drift laminated units. Claystones show extensive bioturbation.

A second laterally continuous intraformational conglomerate outcrops roughly 6 to 8 m below the KBS Tuff. Another similar repetitive sequence of fining-upward sandstone to mudstone strata overlies the intraformational conglomerate.

The final distinctive horizon occurs 2 to 4 m below the KBS Tuff and is restricted to area 105. It is a thin elongate body consisting of a northwest-southeast trending complex of bioclastic and biolithic carbonate. It is divisible into three units, the lowest two carbonates locally become packed with gastropods, with the second bed displaying the better preservation of the two. The capping carbonate is a highly recrystallized packed pelecypod biolithite with numerous tabular cavities (10 cm diameter) on its upper surface. The bed grades and interfingers laterally with fine-grained sandstone and siltstone.

The Lower Member stratum above the intraformational conglomerate marker in area 131 is of measurably coarser grain with lenticular to thin-bedded sandstone being more abundant. In this respect area 131, because of its closer proximity to the basin margin, correlates well with a general north to south fining from the dominant conglomerates and pebbly mudstones of area 129.

The uppermost Lower Member strata are surprisingly uniform. They are composed of a persistent moderate brown (5 YR 3/4) claystone overlain by isolated lenses of KBS Tuff and a capping mudrock and fine-grained sandstone beds.

The KBS Tuff is the key stratigraphic marker horizon of the Lower Member. Occupying depressions on a claystone-mudstone stratum, the tuff outcrops sporadically over the entire study area, locally reaching thicknesses of 3 m. Exposures are usually of very local lateral extent (less than 10 m) but become more continuous in area 131. The tuff in area 105 occupies swales trending northeast-southwest. Where thickest, the tuff is normally a composite of several tuff and tuffaceous layers. At the FxJj 1 archaeological site (area 105), the tuff is light gray (N7) and consists predominantly of fine-grained glass shards. It appears to be massive in structure. This site overlies another tuffaceous bed which contains abundant silt- and sand-sized detritus that is parallel- to ripple-laminated with occasional contorted bedding. Stone artifacts have been found along the

sharp contact between the two beds. Farther south, three
subdivisions are recognizable in the outcrop. The lowest is
a ripple-laminated to small-scale planar cross-bedded and
occasionally contorted white (N9) tuff with thin intercalated
claystone stringers. This lenticular unit is overlain by
ripple-laminated light gray (N7) tuff containing granule- to
cobble-sized pumice clasts. The uppermost unit is a pinkish
gray (5 YR 8/1) tuffaceous siltstone. It is structureless
except for worm burrows.

The KBS Tuff is conformably overlain in ascending order
by a mudstone, a siltstone, and a fine-grained sandstone complex. The mudrocks appear very similar to, but are coarser
than, the claystone-mudstone strata immediately underlying
the tuff. White (N9) calcareous concretions (1-3 cm long)
weather from the rocks both above and immediately below the
KBS Tuff and accumulate on the weathered "popcorn-like" surface of the exposures. The concretions, plus the irregular
fracture and monotonous texture, typify the uppermost mudrocks of the Lower Member. The friable to loose sands above
the tuff, best developed in area 131, appear structureless
and are usually truncated by lenticular, conglomeratic sandstones, the basal content of which marks the boundary between
the Lower and Upper Members. The sandstones occupy erosional
channels into Lower Member strata. A maximum of 10 m of
downcutting can be demonstrated immediately north of FxJj 1.
Here the Upper Member channels have cut through the KBS Tuff
to a level of 6 m below its basal contact.

Upper Member. Principal exposures of the Upper Member
occur on the Karari Escarpment and the Koobi Fora Ridge south
of the Laga Bura Hasuma. Small, isolated outcrops occur in
the Ileret area. The unit attains a thickness of 88 m on the
Koobi Fora Ridge and in the Ileret area but thins to 43 m in
the Karari Escarpment and to just 6 m 50 km to the south near
Shin (Bower, 1974).

Tuffs are numerous in the Upper Member, but only the
uppermost are sufficiently extensive and distinctive for
correlation. In the Ileret, Karari, and Koobi Fora areas,
these uppermost tuffs are the Chari, Karari, and the Koobi
Fora, respectively. The Koobi Fora Tuff complex appears to
be slightly older than the Chari-Karari association (1.57
± 0.00 m.y. B.P. versus 1.32 ± 0.1 m.y. B.P.) (Fitch and
Miller, 1976; Cerling et al., 1975).

The Upper Member outcropping in the scarp face of the
Karari consists of a basal lenticular sandstone and siltstone
sequence beneath a complex of tuff and reworked tuffaceous
sediment, termed the Okote Tuffaceous Siltstone Complex.

Basalt conglomerate and lenticular sandstone occur interbedded with fine sediments of the Okote. This entire
sequence interfingers westward in the Koobi Fora Ridge with
interbedded granule conglomerate, siltstone, bioclastic carbonate, algal stromatolites, and tuffs.

Johnson and Raynolds (1976) measured 155 m of Upper
Member strata in a complexly faulted area adjacent to the
Koobi Fora Ridge. There, cyclic deposits consist dominantly
of mudstones, fine-grained sandstone, and abundant algal
biolithites capped by thin, coarser grained sands.

As redefined by Vondra and Bowen (1978), the basal contact of the Upper Member is drawn at the base of the first
conglomeratic channel sandstone stratigraphically above the
KBS Tuff. The contact records a major lithologic change in
the Koobi Fora sediments that is recognizable throughout the
basin and eliminates boundary confusion where the KBS Tuff is
absent. From the Karari Escarpment westward, the contact
between the members appears to grade from disconformity (10 m
of measurable downcutting) to paraconformity (little or no
erosion observed). The upper limit of the Koobi Fora Formation is the top of the Chari and Karari tuffs.

Guomde Formation

The Guomde Formation is restricted in distribution to
the Chari Ridge where it attains a maximum thickness of 37 m.
Lithologically, it consists of laminated siltstone with
intercalations of thin bioclastic carbonate and lenses of
trough cross-bedded arkose. A prominent, though unnamed,
lenticular tuff occurs approximately in the middle of the
unit. The Guomde disconformably overlies or is in fault contact with the Koobi Fora Formation.

Galana Boi Beds

The status of these thin Holocene lake deposits has been
kept informal due to uncertain lateral relationships (Bowen
and Vondra, 1973). The Galana Boi Beds unconformably overlie
portions of the Guomde, Koobi Fora, and Kubi Algi Formations.
Generally less than 10 m, the beds thicken to 32 m locally on
the Koobi Fora Ridge (Fig. 3). The sediment is principally
gray diatomaceous siltstone, the color contrast being a
distinguishing characteristic. Locally, the beds contain
trough cross-bedded to ripple-laminated subarkose and are
subject to rapid lateral and vertical variation.

The Galana Boi Beds are of interest not only for their
wealth of artifacts and fossils but also for their geographic

and structural occurrence. The beds represent a major Holocene transgression of the lake and have been faulted, along the Kokoi Horst, 120 m above the present lake level. Assigned an age of 9360 $\pm$ 135 m.y. B.P. (Vondra *et al.*, 1971), these lake sediments record major tectonic activity within Recent time.

Lithostratigraphical Analysis of the Koobi Fora Formation

Lithofacies and Depositional Environments

Bowen (1974) and Vondra and Bowen (1976, 1978) describe four major lithofacies, each interpreted to represent a different depositional environment within the East Turkana Basin. All occur within the Koobi Fora Formation and have similarly been detailed for the Karari Escarpment. The lithofacies and corresponding environment are as follows (Vondra and Bowen, 1976):

1) The laminated siltstone facies: prodelta and shallow-shelf lacustrine.

2) The arenaceous bioclastic carbonate facies: littoral lacustrine - beach and barrier beach and associated barrier and supralittoral lagoons.

3) The lenticular fine-grained sandstone and lenticular-bedded siltstone facies: delta plain - distributary channel and interdistributary flood basin.

4) The intertongued, lenticular conglomerate, sandstone and mudstone facies: fluvial channel and flood plain.

The environments are complexly intertongued but generally document a westward regression which partially fills the Turkana half-graben. Several subenvironments can be recognized within each major depositional environment. Lake level determined the distribution of environments within the East Turkana embayment and generally an overall westward progradation of lithofacies through time is documented from the rock record.

Findlater (1976) recognized five sedimentary depositional environments based on modern analogues from the East Turkana area. The environments are as follows:

1) lacustrine low energy

2) lacustrine high energy

3) alluvial delta plain

4) alluvial coastal plain

5) alluvial valley plain.

Obvious similarities exist between the two suites of depositional environments recognized, though the latter does not formally define lithofacies indicative of each environment. In the following discussion, lithofacies interpreted after Vondra and Bowen (1976) are presented. The lithofacies have the unifying characteristics of possessing similar lithology, mineralogy, textural properties, sedimentary structures, color, bed thickness, and areal distribution. Association of distinctive features within each lithofacies permit interpretation of the genesis of that lithofacies. Consideration of the stratigraphic position, contact relationships of the rock units grouped into a lithofacies, and the geometry of the rock bodies confirms the genetic interpretation of said and allows the reconstruction of a sequence of depositional events for the Koobi Fora Formation.

Laminated siltstone facies. Fine-grained, uniform bedding characterizes the laminated siltstone facies throughout the basin. Parallel-laminated to thin-bedded, yellowish gray mudrocks alternate occasionally with thin intercalations of laminated, grayish orange feldspathic litharenites, grayish orange packed molluscan biosparudites, and laminated light gray bentonitic tuffs (Bowen, 1974). The dominant mudrocks contain abundant, plate-shaped limonitic concretions along well-defined bedding planes (Fig. 6). The mudrocks have been measured 18 m thick locally but usually occur in units of 2 to 4 m (Bowen, 1974). Figure 7a illustrates the characteristic appearance of the lithofacies.

As concluded by Bowen (1974) and Vondra and Bowen (1976), the lithofacies of alternating bedding and dominant parallel- to ripple-lamination can be ascribed to proximal prodelta and delta shelf environments of deposition. These are low energy, shallow-shelf lacustrine areas well in front of and lateral to deltaic influx, dominated by fine-grained sedimentation and only occasionally interrupted by current activity which introduces sand and concentrates lacustrine gastropods.

Laminated siltstone facies
(Prodelta and shallow-shelf lacustrine)

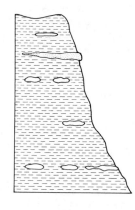

PRIMARY STRUCTURES: parallel lamination

BEDDING: parallel laminated to thinly bedded

FACIES RELATIONSHIP: thickens toward depocenter, interfingers with littoral lacustrine & delta plain deposits

Figure 6.   Depositional characteristics of the laminated siltstone facies.

Figure 7. Photographs of lithofacies

Laminated siltstone facies
a. Typical exposure showing characteristic
    bedding and fracture

Arenaceous bioclastic carbonate facies
b. Gentle lakeward-dipping crossbedding and
    steeply-dipping avalanche cross-beds of a
    sand beach deposit

c. Massive gastropod sandstone

d. & e. Algal-coasted clasts in an intraforma-
         tional conglomerate

Arenaceous bioclastic carbonate facies
(Littoral lacustrine--beach & barrier beach & associated barrier & supralittoral lagoons)

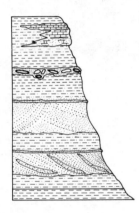

PRIMARY STRUCTURES: *planar & trough x-bedded; ripple drift lamination*

BEDDING: *thin- to thick-bedded*

FACIES RELATIONSHIP: *complexly interfingered between lacustrine & delta deposits*

Figure 8.   Depositional characteristics of the arenaceous bioclastic carbonate facies.

The mudrock mineralogy and texture are uniform, within expected limits, throughout the stratigraphic section of the basin. These mudrocks are predominantly montmorillonite with minor feldspar and quartz and trace accessory clay minerals; kaolinite, illite and calcite. Quantitative estimates of the whole rock mineralogy list montmorillonite, 85%; feldspar, 10%; and quartz, 5%. Secondary selenite veins, vertical to subvertical, 1 cm thick, are common in the laminated silt-stone facies.

Texturally the mean grain size averages within the fine silt to coarse sand range and the beds are poorly to extremely poorly sorted. The central tendency of grain size in sediments is controlled by both the size of the particle available for transport as well as the competence of the transporting medium. The mudrocks reflect the proximity of the surrounding basaltic highlands as well as final deposition in a low competence environment.

The laminated siltstone facies comprises most of the Lower Member along the Koobi Fora Ridge (Bowen, 1974). From here the facies, vertically and laterally along the Karari Escarpment, interfingers with the arenaceous bioclastic carbonate facies. The thin band along the base of the escarpment is unfossiliferous but contains abundant limonitic concretions. These geothite-cemented mudclasts increase in size southward toward area 105 suggesting more stable lacustrine conditions away from the basin margins.

Arenaceous bioclastic carbonate facies. The typical depositional unit in the arenaceous bioclastic carbonate facies grades from a basal parallel- to ripple-laminated feldspathic litharenite and arenaceous bioclastic carbonate capped by a persistent, structureless to ripple-laminated sandstone (Fig. 8). Thickness of each unit may be as much as 8 m.

Figure 7b shows a vertical profile of the feldspathic litharenite exhibiting the distinguishing complex of primary structures. Within the low-angle (4° to 9°), laminated foresets, small- to large-scale trough cross-bedding and ripple-drift-lamination are clearly evident. Truncating the planar sets and dipping steeply (20° to 21°) in the opposite direction are thinner, planar to gently curved foresets of laminated sand. In succession above both sets is large-scale trough cross-bedding capped by a thin 5 cm thick gastropod sandstone. The sandstone sequence is capped by a mudstone drape and laterally interfingers with abruptly truncated mudstones of considerable local dip.

Figure 7c displays the bioclastic nature of the lithofacies. The carbonates consist predominantly of a gastropod
and sand grain framework cemented with calcite. The gastropods are generally recrystallized or have been removed by
solution leaving partial voids surrounding an internal
calcite-cemented, fine-grained sand filling. Faint cross-
bedding is locally present as a continuation of the planar
cross-bedding of the subjacent sandstone.

The stratigraphic succession of the lithologies and primary sedimentary structures in this lithofacies have been
attributed to a beach and barrier beach and associated
barrier and supralittoral lagoon environment (Davies et al.,
1971; Dickinson et al., 1972; Wunderlich, 1972; Reineck and
Singh, 1975; Harms, 1975). This facies consists of three
subfacies: (1) the lower laminated silty sandstone subfacies, (2) the planar and trough cross-bedded sandstone subfacies, and (3) the upper structureless sandstone subfacies.
These represent accumulation in shoreface, foreshore, and
dune subenvironments, respectively (Reineck and Singh, 1975;
Harms, 1975). The highest energy generated by waves in a
beach environment occurs in the breaker zone of the upper
shoreface and swash zone of the foreshore (Harms, 1975). It
is in these zones that the gastropod tests have been concentrated and incorporated with shoreward migrating megaripples, such as the feldspathic litharenites just described.
The thin gastropod-bearing sandstone to massive arenaceous
bioclastic carbonates are interpreted to be the concentrated
lag atop the beach in the swash and upper shoreface zone. As
the megaripples migrate onshore in progressively shallower
subaqueous to subaerial conditions, they accrete to the beach
and finally form a ridge and runnel system in the foreshore
and backshore (supralittoral) zones. Characteristically,
they exhibit two types of stratification, low-angle,
lakeward-dipping cross-bedding and high-angle, generally
shoreward-dipping, avalance cross-bedding. The laterally
interfingering mudstones are remnants of backbeach lagoon
accumulations adjacent to the prograding beach.

As would be expected, the sandstones of the prograding
beach environment show the greatest mineralogical and textural maturity. The quartz-feldspar-rock fragment (QFR)
ratio from sequences in the Karari Escarpment (area 118-105)
average $Q_{41}:F_{25}:R_{34}$ (White, 1976). Quartz grains are sub- to
well-rounded, well-sorted, and exhibit dominantly undulose
extinction (common for quartz sand basin-wide). Feldspar
compositions are mostly orthoclase and sodic through
bytownite plagioclase with a smaller percentage of micro-
cline. Rock fragments preserved in the beach swash zone are
dominantly volcanic (basalt and chert/chalcedony) in origin.

Compositionally then these sands are volcanic arenites, containing a slightly higher quartz fraction than sands in other lithofacies.

The arenaceous bioclastic carbonates are packed biosparudites after the nomenclature of Folk (1974). The framework comprises replaced packed molluscan and ostracod tests, casts, and molds. Those identified in the Koobi Fora include Melanoides sp., Cleopatra sp., and Mutela sp. (Vandamme and Gautier, 1972; Gautier, 1976). The same authors also identified the pelecypods Cobicula sp. and Nyassunia sp. Gastropods are dominant. Poor preservation and almost universal replacement of the tests characterizes most of these carbonates; some in the Chari and Koobi Fora ridge areas are well-preserved and have been extensively studied by Williamson (1978).

The arenaceous carbonates are only end members of a continuum of observed rock types beginning with the more abundant calcareous sandstone lenses. A gradual change to a disrupted detrital grain framework accompanies an increasing calcite content. The orthochemical carbonate may locally exceed 70%. Clay minerals are common in the matrix.

Algal-stromatolite biolithites are also locally abundant in very shallow littoral, supralittoral, and back beach lagoonal subenvironments. Filamentous algae produce a calcareous ooze which coats subaqueous growth surfaces, trapping and binding any detritus coming into contact. Johnson (1974) recognized several hemispheroid, spheroid, and mat forms described by Logan et al., (1964) and Kendall and Skipworth (1968).

The narrow, comparatively thin geometry of the arenaceous bioclastic carbonate facies is complexly interfingered with remaining lithofacies. Units occur throughout the Koobi Fora Formation along western Koobi Fora Ridge (Johnson and Raynolds, 1976) and are restricted to the Lower Member along the Karari Escarpment (Bainbridge, 1976; White, 1976).

Two sandy prograding beach profiles outcrop in area 105 but are not as well-developed near the basin margin in area 130-131. In addition, the facies is represented by two persistent intraformational conglomerates generally less than 0.2 m thick although depths exceeding one m are recorded locally. The erosional basal contact, abundance of sedimentary intraclasts, and the thin but laterally persistent nature of the conglomerates support the transgressive shoreline hypothesis (Bowen, 1974; Johnson, 1974). The thinness and lateral continuity are characteristic of transgressive

beach sequences versus the much thicker but less persistent
regressive successions (Walker and Harms, 1971).

Of particular interest is the depositional history
interpreted for the lower conglomerate. In locations where
the clasts are dominantly cobble-sized, a significant percentage is algal encrusted and imbricated within the fossiliferous, arenaceous calcite matrix (Fig. 7d & e). The occurrence of whole and fragmented, coated and uncoated intraclasts suggests a complex origin. To expose the mudstone
concretions of the immediately subjacent strata, it is
suggested that an unknown thickness of the mudstone was
removed by erosion accompanying transgression of the lake.
After becoming clasts, the nodules must then have been deposited in a quiet lagoon-type environment where algae could
provide a means for stromatolithic encrustation. Johnson
(1974) observed similar encrustations near Koobi Fora Spit.
Classified as oncolites (Logan et al., 1964), the clasts
exhibit a coating development from less than 1 mm to complete
spherical stromatolites with a 2 cm carbonate overgrowth.
The oncolites have been observed forming a subaqueous shoal
environment characterized by sufficient current and wave
activity to periodically mix the clasts (Logan et al., 1964).
Apparently the thickness of coating can be attributed to
residence time in the algal environment and strength of wave
and current activity. A very few coatings of the associated
invertebrate and disarticulated vertebrate fragments have
also been observed. The final conglomerate of coated and
uncoated clasts may have been the result of high energy wave
action during storm-driven tides or renewed transgression of
the lake due to unique tectonic events within the rift.

Findlater (1976) defines the base of this short-lived
transgressive bed as the post-Tulu Bor erosion surface. The
magnitude of lake level change sufficient to have created
these beds may not have been more than a few meters assuming
the alluvial plain near the shoreline was broad and of very
low relief. The horizon does not represent the advent of an
erosion cycle as the disconformity between the Upper and
Lower Members and should be considered as an essentially
isochronous time line much like the tuff marker beds.

Lenticular fine-grained sandstone and lenticular-bedded
siltstone facies. Consistent with the entire East Turkana
Basin, strata of the lenticular fine-grained sandstone and
lenticular-bedded siltstone facies dominate the Lower Member
along the Karari Escarpment. The lithofacies intertongues
with the arenaceous bioclastic carbonate facies at the base
and underlies the lenticular conglomerate, sandstone, and
mudstone facies.

Within the facies, variations in texture and bedding types are noted as shown in Figure 9. Lenticular bedding with thick isolated lenses (Reineck and Singh, 1975) is common. Mudrocks are finely parallel-laminated to thin-bedded. Isolated sandstone lenses generally exhibit small-scale laminated cross-bedding. The mudrocks appear as massive, structureless, fining-upward, rhythmic sequences but occasionally display faint parallel lamination. These exposures differ from the laminated siltstone facies by their lack of thin, distinct bedding.

Lenticular channel complexes of cross-bedded sandstone, interbedded sandstone, and siltstone occur frequently overlying the rhythmic mudstones (Fig. 10a-d). The sandstone displays large planar cross-beds, small-scale trough and ripple (Kappa Type I) cross-bedding, and ripple lamination. Thickness of the sandstone may exceed 25 m (Bowen, 1974). The bulk of the interfingering complex is a sequence of repetitive fining-upward rhythms of fine-grained sandstone to claystone (Fig. 10a and b).

The distinctive bedding of the lenticular fine-grained sandstone and lenticular-bedded siltstone facies is indicative of sedimentation on a delta plain (Bowen, 1974). The delta plain can in turn be differentiated into several subenvironments, namely the upper distributary channel flanked by subaerial distributary levee and flood basin zones and the subaqueous delta front zones including the distributary channel, subaqueous levee, distributary mouth bar, and distal bar. The adjacent but deeper water, prodelta and shallow shelf lacustrine environments are represented by the laminated siltstone facies.

The bedding types observed in this lithofacies are not indicative of any one specific subenvironment, but from the total assemblage of primary structures present (Coleman and Gagliana, 1965), it is possible to delimit probable depositional environments. For instance, abundant mudrocks and extensive bioturbation are most likely to be found on subaerial levees and floodplains only seasonally inundated by floodwater. The greatest variety of sedimentary structures occur within the delta front and subaerial levee sediments. The remaining subenvironments frequently display monotonous thin-bedded and often rhythmic repetitions possibly resulting from intermittent high-water surges and subsequent slackwater deposition. Such fining-upward rhythms are well-developed on subaqueous levees and distal bars and are often pronounced in other subenvironments. At the base of the lithofacies, the lenticular-bedded mudstone with thick sand lenses represents part of the subaqueous delta front. The bedding is the

Lenticular fine-grained sandstone and flaser-bedded siltstone facies
(Delta Plain--distributary channel and interdistributary flood basin)

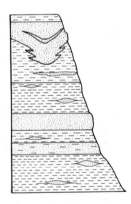

PRIMARY STRUCTURES: parallel to ripple
lamination; planar & trough x-bedding in channel

BEDDING: rhythmic; laminated to thinly bedded

FACIES RELATIONSHIP: constitutes greatest volume
of deposits; interfingers extensively with other
facies

Figure 9.   Depositional characteristics of the fine-grained
sandstone and flaser (lenticular-bedded) siltstone
facies.

Figure 10.   Photographs of lithofacies

Lenticular fine-grained sandstone and
lenticular-bedded siltstone facies:
a.   Exposures of the rhythmically-bedded strata
b.   Ripple-laminated sandstone lens
c. & d.   Distributary channel and laterally equivalent rhythmic strata

Lenticular, conglomerate, sandstone, and mudstone facies:
e.   Trough-shaped KBS Tuff deposit within fine-
grained flood plain sediments

Lenticular conglomerate, sandstone, and mudstone facie
(Fluvial channel and flood plain)

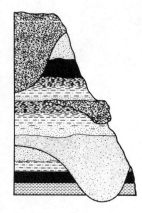

PRIMARY STRUCTURES: large and small scale x-stratification & lamination

BEDDING: thinly laminated to thick-bedded

FACIES RELATIONSHIP: thickest along basin margin; interfingers with and overlies delta plain deposits

Figure 11.   Depositional characteristics of the lenticular conglomerate, sandstone, and mudstone facies.

result of dominantly suspended-load sedimentation interrupted by infrequent bed-load surges of current or wave activity. The repetitive fine-grained cycles above closely resemble the subaqueous levee and distal bar characteristic bedding of Coleman and Gagliana (1965). The stratigraphically higher interfingering complex represents the distributary channel, proximal subaerial levee, and flood basin sedimentation displaying mudcracking and coarser cyclic repetitions. That the complex is a near-shore sequence is evidenced by the interfingering bioclastic carbonate, interpreted as a beach deposit (upper Lower Member, area 105).

Similarly, deltaic sedimentation dominates the Lower Member in area 131. The generally coarser-grained strata may be a reflection of either closer proximity to source area or greater influence of upper deltaic distributary channel, subaerial levee, and floodplain subenvironments. Vertically within this lithofacies along the escarpment, the trend is toward increasingly shallow subaqueous to subaerial deltaic environments.

Polymictic channel lag conglomerates consisting of volcanic, metamorphic, and plutonic igneous rock fragments occur infrequently at the base of the distributary channels (Bowen, 1974). The composition of the sands ranges from volcanic arenite to lithic arkoses with lower quartz:feldspar and quartz:rock fragment ratios than winnowed beach sands. Texturally, mean grain size is similar but sorting and grain shape are generally less mature.

Intertongued, lenticular conglomerate, sandstone, and mudstone facies. Conformably overlying the lenticular finegrained sandstone and lenticular-bedded siltstone facies capping regressive cycles in many areas is the intertongued lenticular conglomerate, sandstone, and mudstone facies. The composite depositional unit of the facies comprises an erosionally-based, basal conglomerate, followed by coarse- to fine-grained arkose or feldspathic litharenite and capped by and laterally-equivalent to fine-grained mudrocks and tuffs (Bowen, 1974) (Fig. 11). Conglomerate lenses are prevalent along basin margins and reflect oligomictic to polymictic rock components. Sandstones display a similar sequence of primary structures as the deltaic distributary channels as well as sinuour outcrop geometry. Arenaceous siltstones and claystones contain abundant calcareous root casts and irregularly-shaped concretions. Many variations are observed in these sedimentological parameters making the lithofacies the most heterogeneous. Part II discusses many of these subfacies and variations recognized in the Upper Member, Karari Escarpment.

A fluvial depositional environment is proposed for these uppermost sediments. Within the fluvial environment, the stream systems are dominantly either braided or meandering. Of the many factors that influence braiding versus meandering the factors of slope and sediment grain size play significant roles. Generally the stream becomes more braided and less sinuous with increasing gradient and coarser grain size (Leopold and Wolman, 1957; Schumm, 1968; Coleman, 1969; Williams and Rust, 1969; Steel, 1974; Walker, 1975).

Flood plain sediments can be associated with several distinct subenvironments, namely point bars, channel fills, crevasse-splays, natural levees and flood basins. These represent both lateral and vertical accretion on flood plains. Lateral migration of the fluvial channel involves primarily point bar and lesser channel lag deposition while vertical accretion results from suspended-load sedimentation under lower energy crevasse-splay, natural levee, and flood basin conditions. Vertical accretion deposits commonly overlie the normally dominant lateral accretion deposits (Allen, 1965).

To interpret the history of flood plain development, it is important to realize the relative proportion of channel and overbank deposits. Leeder (1978) catalogues the controlling factors as channel type and magnitude, rates of lateral migration, avulsion, and vertical flood plain accretion. Each of these factors in turn depends on sediment load, discharge, gradient, tectonics, and climate. In particular factors influencing vertical accretion rates include channel size, flood plain width, flood sediment concentrations, overbank flood periodicity, and others (Leeder, 1978).

Fluvial environments are inferred for the uppermost Lower Member and Upper Member along the Karari Escarpment, the Upper Member along the eastern Koobi Fora Ridge toward Shin, the Upper Koobi Fora east of Ileret, and between the Kokoi Horst and Koobi Fora Ridge (Bowen, 1974). Archaeological and anthropological interest in the Karari area has again stimulated detailed documentation of the lithofacies' distribution. In area 129 Bainbridge (1976) subdivided the Lower Member lithofacies into the basalt conglomerate and pebbly mudstone subfacies representative of alluvial fan deposits and the lenticular sandstone and mudstone subfacies, the fluvial channel, and flood basin deposits.

The basalt conglomerate and pebbly mudstone subfacies. This subfacies consists of basalt pebble to boulder conglomerate whose sorting varies from poor- to well-sorted. Inter-

bedded with the conglomerates are finer grained sediments
which have distinctive internal features.

The conglomerates are composed predominantly of basalt
clasts with some chert and feldspar fragments and rare granite pebbles. The thickness of the conglomerates decreases
upward through the subfacies from a maximum of 15.6 m to
thinner units typically 0.5 m thick. Sorting, on the other
hand, increases upward; thinner upper units are well-sorted.
The units coarsen upward; a single outcrop may be composed of
several units each of which displays a coarsening-upward
sequence. Clasts within the conglomerate are crudely imbricated and show some horizonation which is better developed in
stratigraphically higher units. The upper conglomerates have
sharp, irregular, or very weakly erosional basal contacts and
horizontal upper contacts resulting in units of sheet-like
dimensions. Laterally the graded bedding in the upper conglomerates gives way to planar and low angle trough type
cross-bedding. These cross-bedded lenses have a channel
morphology in cross-sectional view and erosional contacts
with the surrounding conglomerate. The channels average 1.5
m in thickness. The upper conglomerates often contain gastropod and oyster shells at their base. The geometry of the
basal conglomerate is lenticular; large-scale channel-fill
cross-bedding is typical.

Pebbly mudstones are gradationally interbedded with the
sheet conglomerates. They consist of basalt clasts ranging
in size from sand to cobbles dispersed in a supporting matrix
of mudstone. Limonite, calcite, and gypsum veins and halite
crystals occur in the mudstone. The mudstones are interpreted to have been deposited in mudflows.

Mudflows are mass movement phenomena of varying viscosity having the competence to transport a wide range of particle sizes. The group is an intermediate series of a gradational continuum of mass movement phenomena characterized by
varying proportions of water, clay, and rock debris. With
increasing viscosity, they tend toward earthflows, and with
decreasing viscosity, toward water-laid deposits. Mudflows
may be recognized by abrupt, well-defined margins, by a
lobate (tongue) geometry, and by the lack of any graded
bedding or particle orientation (Bull, 1964). Favorable
conditions for mudflow formation are the presence of
unconsolidated material which contains sufficient clay to
make the mass less cohesive when wet, steep slopes, short
periods of abundant rainfall, and sparse vegetative cover.

The subfacies is interpreted to be alluvial fan deposits

derived from the Suregei Cuesta. Mudflow deposition explains the characteristics of the pebbly mudstones such as the lack of particle orientation as well as the wide range of clast sizes present within them. The large amounts of montmorillonite present may be responsible for the inability to distinguish lobate geometries in the mudstone units. Churning by the clay could have obliterated such morphologies.

The sheet deposits are water-laid and are formed by a network of shallow (20 cm or less) anastomosing braided streams which fill their channels in surges and then shift a short distance away. Channel deposits formed seasonally on the alluvial fans (Bull, 1964). The presence of gastropods and oyster beds at the base of the sheet units indicates that these were deposited into lake waters near basin margins.

The lenticular sandstone and mudstone subfacies. This subfacies is present at the top of the Lower Member from southern area 129 to area 105 in a complex up to 10 m thick. The base, most commonly occurring lithology is a moderate brown (5 YR 3/4) claystone to the south. The claystone has an irregular fracture, appears structureless, and contains abundant white (N9) calcareous concretions, 1 to 3 cm long. Northward the claystone interfingers with a lenticular feldspathic litharenite. The KBS Tuff stratigraphically overlies the claystone and occupies trough-shaped depressions (Fig. 10e). The tuff is relatively pure to highly argillaceous and is parallel- to ripple-laminated. Calcified root casts as well as pumice fragments are locally common. Where thickest, the tuff can be divided into two or more subunits, each having different lithology, structure, and color. Above the tuff occur more very poorly sorted mudrocks and fine-grained sandstones, approximately 2 to 4 m thick; faint parallel to ripple lamination and small-scale cross-bedding are common. These sediments do not exhibit the characteristic point bar cross-bedding sequence of the meandering stream depositional model (Allen, 1970; Walker, 1975). Accordingly, the strata may either represent lateral flood basin or natural levee deposits of a very low gradient and sinuous meandering stream system.

Finally, mention should be made of the geomorphic position of the KBS Tuff. That position is interpreted to be sediment traps on the clay floodplains of a meandering stream system. The traps could have been swales on point bars deposits or meander cut-off chutes not yet filled. The aeolian and reworked tuff was deposited into these trough-shaped channels, possibly during flood stage, and protected from subsequent erosion by later flood surges.

Age and Correlation

Thus far in this chapter, evidence for basinal stratigraphy physical correlation and lithofacies for the Koobi
Fora Formation has been based on extensive stratigraphic
field studies and is very well documented. It would be
extremely fortuitous for all the varied studies of stratigraphy, paleontology, paleomagnetism, and radiometric dating
to fit one correlation scheme intrabasinally with no discrepancies. However, differences in data or the interpretation
of that data have resulted in a number of inconsistencies
emerging from application of the above subdisciplines to
problems of dating in the East Turkana Basin sedimentary
sequence. The scope of these geochronology discrepancies
have been reviewed in Vondra, Burggraf, and White (1978) and
Behrensmeyer (1978) and will only briefly be summarized here.

Radiometric $^{40}Ar/^{39}Ar$ techniques (Fitch and Miller,
1976; Fitch et al., 1976) date the Koobi Fora Tuff markers as
follows: (1) Chari Tuff: 1.28 ± 0.23 m.y. B.P., (2) Karari
Tuff: 1.23 ± 0.1 m.y. B.P., (3) Koobi Fora Tuff: 1.57 ± 0.0
m.y. B.P., (4) Okote Tuff: 1.56 ± 0.02 m.y. B.P., (5) KBS
Tuff: 2.42 ± 0.01 m.y. B.P., and (6) Tulu Bor Tuff: 3.18
± 0.09 m.y. B.P. Hurford et al., (1976) independently determined a KBS age of 2.44 ± 0.08 m.y. B.P. based upon fission
track dating. Curtis et al. (1975) proposed two apparent
ages for the KBS Tuff (1.82 ± 0.004 m.y. B.P., area 131; 1.60
± 0.05 m.y. B.P., areas 105 and 10) proposing the existence
of two KBS tuffs within the Karari Escarpment. Suid fauna
evolutionary trends (Cooke, 1976; Harris and White, 1977) are
interpreted to also suggest different stratigraphic levels of
the KBS Tuff along the escarpment. There is, however, conclusive field evidence, as mentioned earlier in this chapter,
against this, requiring the difference in absolute age or
faunal trends to be explained some other way.

The calibration of observed magnetozones (Brock and
Isaac, 1974; Hillhouse et al., 1977) is also linked to the
age of the KBS Tuff. The older date of 2.42 m.y. B.P. per-
mits extensive basinal uplift plus climatic changes and
allows erosion sufficient time to create the Upper-Lower
Member disconformity and inferred environmental changes (Part
II).

Interbasinal correlations within the Turkana Depression
for the several Plio-Pleistocene localities (Fig. 1) have
been proposed by several authors (Behrensmeyer, 1976; Harris
and White, 1977; Cerling et al., 1979). As no stratigraphic
continuity exists between these isolated remnants, inferred
correlations are based upon faunal, radiometric, and geo-

chemical trends.   Physical correlation by similarity of the
respective   stratigraphic   sequences   is   precluded   either
because the sections are known to be diachronous via radiometric and/or faunal evidence or because the respective
depocenters, if consanguineous, occupied various positions
relative to the rift axis.   The latter situation governs the
overlapping ages of the Omo and East Turkana basins.   Similar
faunal trends have been inferred (Harris and White, 1977),
but absolute dating techniques have not been able to resolve
the dilemma.   Cerling et al. (1979) believe that chemical
analyses of volcanic glasses fingerprint three tuffaceous
horizons in both the Shungura Formation of the Omo and the
Koobi Formation: namely, (1) Tuff L with the Chari/Karari
tuffs; (2) Tuff H2 with the KBS Tuff; and (3) Tuff H4 with a
tuff at the stratigraphic level of the KBS, eastern area 105.
Further resolution, if possible, will undoubtedly stimulate
inferences regarding the hominid habitat beyond now-isolated
rift basins.

Provenance and Paleogeography

Sedimentation patterns in the East Turkana embayment
were influenced by sediment supply and local, intrabasinal
subsidence.   Sediment supply is a function of provenance
geology and relief, regional tectonism, climatic conditions
and vegetative cover.   Subsidence is reflected by changes in
lake level, local relief, and subsequent migration of depositional environments (shoreline shift).   The lithofacies
distribution of the Koobi Fora Formation outline irregular
north-south belts recording an overall regression (progradation) of the environments gradually infilling the embayment.
Sediment supply obviously exceeded subsidence though numerous
transgressive sequences interrupt westward shoreline retreat
during the Plio-Pleistocene.

Two distinct provenances have contributed detritus to
the East Turkana sediments.   The most distant is the Precambrian basement complex of the Ethiopian highland and the
hills flanking Lake Chew Bahir to the northeast and eastern
Omo Valley to the north.   Though the age is uncertain, the
crystalline rocks are principally granitic gneiss and schist
(Walsh and Dodson, 1969).   Mathisen (1977) identified the
heavy mineral suite of four basement samples collected by Ron
Watkins (Birkbeck College, University of London) north of
Lake Chew Bahir and found the common species of biotite, gar-
net, magnetite, apatite, hornblende, and zircon.   Mathisen's
intensive study of the heavy mineral suite of East Turkana
sandstones revealed a dominant plutonic/metamorphic source
for many fluvial and distributary channels and beach profiles.

The volcanic highlands immediately bordering the half-graben constitute the second provenance. Here the rocks consist predominantly of Pliocene flood basalt and alkali olivine basalt, interfingering with minor tuff and sedimentary deposits, overlying Miocene basalt and siliceous ignimbrite flows (Ron Watkins, Birkbeck College, personal communication, 1976). Olivine, a common accessory in the basalt, does not survive in the basin sediments as do hornblende and biotite. Since mudrocks dominate basinal stratigraphy, the volcanic province has been of major influence. Streams draining the volcanic highlands readily brought detritus into the basin. Coarser basement-weathering products were introduced through streams heading to the east and northeast, especially in the present Karari Escarpment area. Evidence for a proto-Ol Bakate gap inlet are inconclusive though the physiographic configuration may have been similar. Longshore drift from the Omo area may also have contributed detritus.

The degree to which each provenance supplied detrital material differs both vertically and laterally within the basin. Certainly the source for the high montmorillonite content, averaging over 80% in the dominant mudrocks, is weather basalt (Deer, Howie, and Zussman, 1966). In addition, the mudrocks analyzed contained trace components of illite, kaolinite, phlogopite-biotite, muscovite, hornblende, zeolite (clinoptilolite), and calcite. The minor clay minerals are alteration products of each suite of rocks; zeolite and carbonates are diagenetic. The presence of the hydrous aluminosilicate, clinoptilolite, is especially indicative of the volcanic source material and the alkaline environment of deposition (Hay, 1966, 1976). The proximal volcanics appear to be a very dominant source of materials constituting the Lower Member. Crystalline detritus is concentrated in the coarser-grained strata; the heavy mineral suites of the plutonic and volcanic provinces and their basin-wide distribution have been estimated by Mathisen (1977).

Differences between fluvial and deltaic environments are not recognizable mineralogically. However, the differentiation of predominantly volcanic material in the floodbasin and shallow-shelf mudrocks, versus the coarser crystalline fraction in the distributary and fluvial channels, is attributable to two factors: (1) the original particle grain size eroded from the source terrain and (2) intraenvironmental sorting processes. It is inferred that the detrital fraction entering the basin from the crystalling highlands supplied the coarser bed load settling very proximal to the channel, leaving only the finest feldspar quartz and the clay minerals to be deposited in alluvial and interdistributary flood basins.

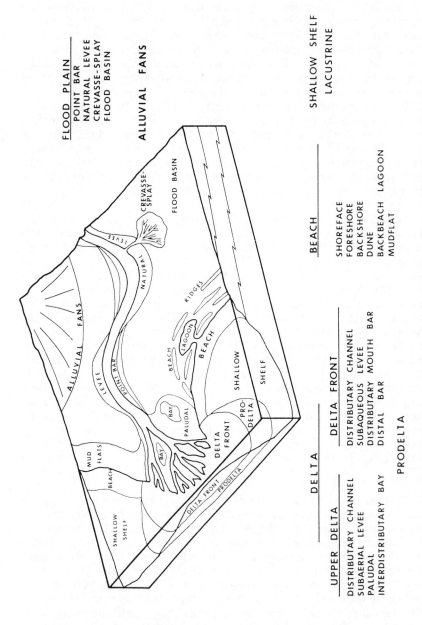

FLOOD PLAIN
POINT BAR
NATURAL LEVEE
CREVASSE-SPLAY
FLOOD BASIN

ALLUVIAL FANS

SHALLOW SHELF
LACUSTRINE

BEACH

SHOREFACE
FORESHORE
BACKSHORE
DUNE
BACKBEACH LAGOON
MUDFLAT

DELTA

UPPER DELTA

DISTRIBUTARY CHANNEL
SUBAERIAL LEVEE
PALUDAL
INTERDISTRIBUTARY BAY

DELTA FRONT

DISTRIBUTARY CHANNEL
SUBAQUEOUS LEVEE
DISTRIBUTARY MOUTH BAR
DISTAL BAR

PRODELTA

Figure 12. Generalized scheme of prograding depositional environments

The variables defining the boundary conditions of the hydrodynamic flow regime include discharge, depth, slope, and particle size and shape, specific gravity of grains, and density and viscosity of the water-sediment mixture (Harms and Fahnestock, 1965). Evidence suggests that within the basin, discharge, water depth, and slope were each low throughout the Lower Member interval but locally higher marginal to the bounding highlands during the entire Plio-Pleistocene history of deposition, especially along the Karari Escarpment. It is clear that conditions are overall low energy, interpreted to be within the lower flow regime (Simons and Richardson, 1961; Harms and Fahnestock, 1965).

Figure 12 illustrates the hypothetical depositional scheme conceived for the generally regressive regime. When constructing paleogeographic interpretations, the specific environments cannot be considered as totally discrete units, rather each is part of a continuum in which an almost limitless number of variations and intergradations are possible. This continuum is characterized by vertical gradation and lateral interfingering. The diagram displays a prograding fluvio-lacustrine complex incorporating deltaic and beach environments and is patterned after similar schematics by Coleman and Gagliano (1965), Allen (1970), Kanes (1970), and LeBlanc (1972). The hydrodynamic and geographic parameters suggested by the interpreted flow regime are represented in the generalized diagram. The reader can thus visualize the important continuity between environments of deposition.

Some features of the paleogeography of the area have been previously described by Vondra and Bowen (1976), Bowen (1974), and Findlater (1976). The initial Cenozoic sediments of the Kubi Algi Formation, as discussed earlier, were deposited as localized alluvial fans of detritus shed from volcanic uplands and deposited relatively close to their source. This implies that a coherent drainage network had not yet developed in the basin. During most of the late Pliocene and Pleistocene, the northeastern part of the Turkana depression was inundated by a large embayment which developed initially in the Allia Bay area and subsequently progressed northward (Vondra and Bowen, 1976; Findlater, 1976). However, the lowermost Koobi Fora sediments (i.e., those below the Tulu Bor Tuff) record a widespread regression of the lake beginning in the Allia Bay area prior to the deposition of the Suregei (diatomite) (Bowen, 1974). Computer-generated depocenter simulations have been presented by Findlater (1976).

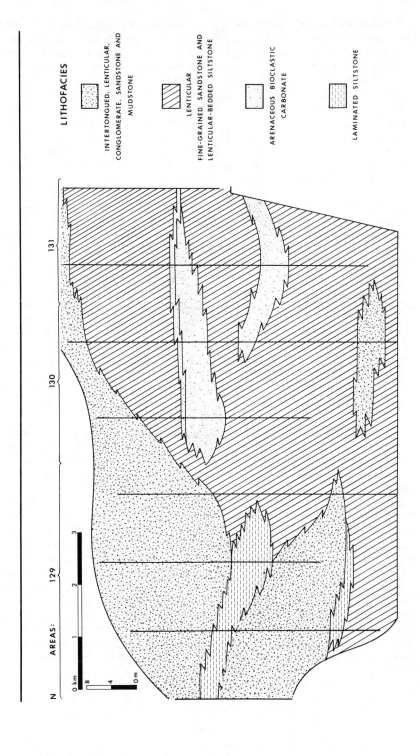

LITHOFACIES

INTERTONGUED, LENTICULAR, CONGLOMERATE, SANDSTONE AND MUDSTONE

LENTICULAR FINE-GRAINED SANDSTONE AND LENTICULAR-BEDDED SILTSTONE

ARENACEOUS BIOCLASTIC CARBONATE

LAMINATED SILTSTONE

AREAS: 129 130 131

N

0 km 1 2 3

8 4 0 m

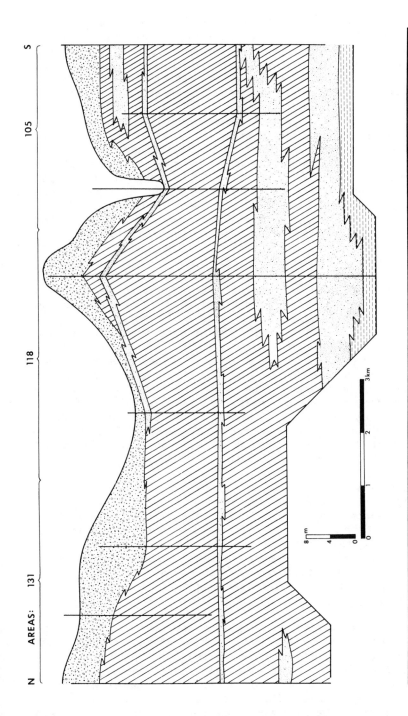

Figure 13. Lower Member lithofacies distribution along the Karari Escarpment. The intraformational conglomerate (IFC) may be used as a reference datum plane.

The paleogeography of the Koobi Fora Formation interpreted from the exposures of the Karari Escarpment - Koobi
Fora Ridge area is discussed here principally in terms of the
drainage present during the deposition of the formation.
Following the initial shedding of detritus into isolated
basins by the monoclinal flexure of the rift to the east, a
structurally controlled, coherent drainage network was developed. Structural control of the drainage resulted in a
configuration essentially similar to that of the present,
i.e., having a north-south orientation turning to the lake
wherever local topography allowed. The main drainage entered
the basin near where it does presently through the 01 Bakate
Gap (Findlater, 1976). Subsidiary minor drainages were
established in the Ileret and northern part of the Karari
areas. Many features of the Koobi Fora Formation at the
Karari Escarpment can be used as evidence that the paleo-
drainage had such a configuration. The sediments in the
formation show a decrease in grain size from the escarpment
to Koobi Fora Spit. This is consistent with depositional
progradation by drainage having a configuration similar to
present day. Facies above and below the KBS Tuff at the
escarpment and to the southwest also argue that this is the
case. Sediments below the tuff at the escarpment are deltaic
plain deposits contrasting with the more lacustrine deposits
found about the tuff at Koobi Fora. Deposits carried into
the area from the north had built a subaerial delta in the
more northerly escarpment area while the Koobi Fora area was
still subaqueous (shallow-shelf lacustrine). The arenaceous
bioclastic carbonate is thicker and possesses better-
developed structures to the south and west. This is consistent with the greater amount of sand and energy believed
present in the environment due to the drainage configuration.

Changes within the facies along the Karari Escarpment
(Fig. 13) can be used to demonstrate the presence of subsidiary drainage in areas 129-130. The silt-clay lamination
present in the laminated siltstone facies in area 129
suggests that the area was proximal to a stream while the
color lamination in the same facies in area 131 was more
distal to current activity. The fine micaceous sands interbedded with laminated siltstones suggest the presence of a
minor stream.

That this subsidiary drainage was structurally controlled can be seen by its continued re-establishment through
time. A large perennial stream flowed through the area in
pre-Tulu Bor time. Its course, using cross-bedding orientation, was similar to that of the sand-bed ephemeral stream
presently in the area. The perennial nature of the stream
can be gauged by the size and abundance of freshwater _Etheria_

sp. populations in the channel. A topographic low was present in the area subsequent to the deposition of the laminated siltstone facies. Arenaceous bioclastic carbonates in area 130, and the previously mentioned delta deposits in area 129 are at stratigraphic levels where equivalent delta plain deposition was occurring in area 131. Such transgression could occur on a delta lobe if sediment supply to the lobe was curtailed (C.D. Masters, oral communication to R. Bainbridge, 1976). Large-scale fluvial activity was established again in late Lower Member times. Large channels cut into Lower Member sediments prior to Upper Member depositon. Again, current direction approximated that presently developed in the area.

The oldest sediments to the south in areas 118 and 105 show gradual lateral as well as vertical transition from shallow-shelf lacustrine, through prodelta, to proximal delta front environments. These delta front subenvironments, in turn, give rise vertically to various beach subenvironments represented by the arenaceous bioclastic carbonate facies. The sequence represents the first of two cycles of prograding deltas capped by shoreline facies.

After an apparent lake transgression above the beach strata, a very general regression from delta front to subaerial distributary channel subenvironments of the lenticular fine-grained sandstone and lenticular-bedded siltstone facies again dominates the section. This sequence is interrupted by two thin transgressive shoreline deposits, the intraformational conglomerates. As suggested previously, these units may represent short-lived storm tide or flood accumulations either on the beach or along broad shoreline mudflats. Whatever the mechanism of formation, the uniformity, lateral persistence, and stratigraphic composition of the conglomerates support a shoreline development hypothesis.

The shallow delta facies also implies that at no time during this interval was the depositional environment very far from lake level. The distributary channel complex below the KBS along the escarpment is a well-defined example of the subaerial-subaqueous delta conditions. The upper delta plain complex, as envisioned in Figure 14, migrates southwestward extending toward the basinal depocenter during Lower Member time in the Koobi Fora area (Vondra, Burggraf, and White, 1978).

The distinctive moderate brown claystone marks the beginning of the flood plain association controlling the remaining Lower Member deposition. The predominantly fine-grained strata are interpreted to include vertical accretion

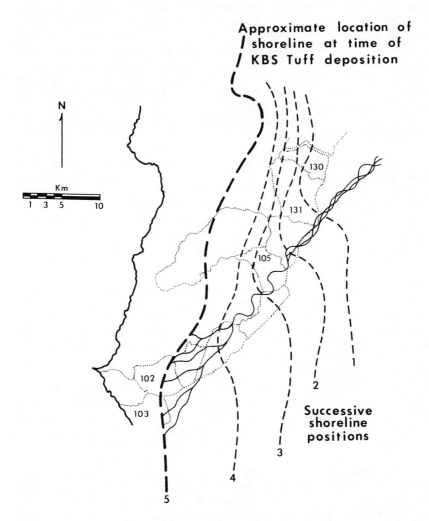

Figure 14.   Schematic representation of shoreline positions
during delta progradation, just after time of
Tulu Bor Tuff deposition to time of KBS   Tuff
deposition (reprinted from Vondra, Burggraf, and
White (1978), Fig. 7 by permission.)

sediments of natural levee and flood basin origin. The KBS Tuff occupied depositional lows on the entire regime until partially eroded. Preservation is restricted to these troughs. The depositional regime abruptly changed with initiation of pre-Upper Member erosion and the deposition of a coarse channel sandstone. Bowen (1974) attributes this change mostly to tectonic downwarping of the basin and/or further uplift and tilting of the cuesta to the east. Paleogeography during Upper Member times is summarized in Part II.

A maximum of 60 m of sediment exists between the Tulu Bor and the KBS Tuff in area 129 (Bainbridge, 1976). The average sediment accumulation rates, based on a 3.18 ± 0.09 m.y. B.P. Tulu Bor Tuff age and the conflicting ages of 2.61 ± 0.26 m.y. B.P. (Fitch and Miller, 1976) and 1.60 ± 0.05 m.y. B.P. (Curtis et al., 1975, area 105) for the KBS Tuff, is 10 and 4 cm/1000 yr., respectively. Behrensmeyer (1974) estimated the Upper Member sedimentation rate at Ileret to be 11 cm/1000 yr. while stating an overall rift valley rate at between 1 and 36 cm/1000 yr. As one would expect, the highest rates of accumulation occur along the rift axis as with the Shungura Formation of the Omo Basin. The 4 and 10 cm/1000 yr. rates presented here assume constant sediment accumulation which neglects such significant variables as differential intraformational erosion and sedimentation, tectonic activity, and diagenetic compaction. Nevertheless, the rates do support the proposed hypothesis of low energy fluvio-lacustrine accumulation during the deposition of the Koobi Fora Formation.

Acknowledgements

Appreciation is extented to the AAAS for sponsoring the symposium entitled "Environmental Setting of Early Man Sites" and encouraging the subsequent monograph of worldwide research.

Iowa State's participation with the Koobi Fora Research Project is sponsored by the National Museums of Kenya and the Lewis Leakey Memorial Institute for African Prehistory and supported by National Science Foundation grants GA-25684, GS-37814, and EAR-7902800. Thanks to Mr. Richard E. Leakey, Director of the Museums, and Dr. Glyn L. Isaac, University of California at Berkeley, for our continued association.

References

Acuff, H.N., 1976, Late Cenozoic sedimentation in the Allia Bay Area, East Rudolf (Turkana) Basin, Kenya (Ph.D. dissertation): Iowa State University, Ames, Iowa.

Allen, J.R.L., 1965, A review of the origin and characteristics of recent alluvial sediments: Sedimentology, v. 5, p. 89-191.

_____, 1970, Physical processes of sedimentation, George Allen and Unwin Ltd., London.

Arambourg, C., 1935, Esquisse geologique de la bordure occidentale du Lac Rudolphe, Mission Scientifique de l'Omo, 1923-1933: Bull. Mus. His. Nat., Paris, fasc. 1.

Bainbridge, R.B., 1976, Stratigraphy of the Lower Member, Koobi Fora Formation, northern Karari Escarpment, East Turkana Basin, Kenya (Master of Science Thesis): Iowa State University, Ames, Iowa.

Baker, B.H., 1970, The structural pattern of the Afro-Arabian rift system in relation to plate tectonics: Royal Soc. London Philos. Trans., Ser. A, v. 267, p. 383-391.

Baker, B.H., and Wohlenberg, J., 1971, Structure and evaluation of the Kenya rift valley: Nature, v. 229, p. 538-542.

Baker, B.H., Mohr, P.A., and Williams, L.A.J., 1972, Geology of the eastern rift system of Africa: Geol. Soc. Amer. Spec. Paper 136.

Behrensmeyer, A.K., 1974, Late Cenozoic sedimentation in the Lake Rudolf Basin, Kenya: Annals Geol. Surv. Egypt, v. 4, p. 287-306.

_____, 1976, Lothogam Hill, Kanapoi, and Ekora: A general summary of stratigraphy and faunas, in Coppens, Y., Howell, F.C., Isaac, G.L., and Leakey, R.E.F., eds., Earliest man and environments in the Lake Rudolf Basin: University of Chicago Press, Chicago, p. 163-170.

_____, 1978, Correlation of Plio-Pleistocene sequences in the northern Lake Turkana Basin - a summary of evidence and issues, in Bishop, W.W., ed., Background to early man in Africa: Geol. Soc. London, Spec. Publ. 5, p. 421-440.

Bowen, B.E., 1974, The geology of the Upper Cenozoic sediments in the East Rudolf embayment of the Lake Rudolf basin, Kenya (Ph.D. dissertation): Iowa State University, Ames, Iowa.

Bowen, B.E., and Vondra, C.F., 1973, Stratigraphical relationships of Plio-Pleistocene deposits, East Rudolf, Kenya: Nature, v. 242, p. 391-393.

Brock, A., and Isaac, G.L., 1974, Paleomagnetic stratigraphy and chronology of the hominid-bearing sediments east of Lake Rudolf, Kenya: Nature, v. 247, p. 344-348.

Bull, W.B., 1972, Recognition of alluvial-fan deposits in the stratigraphic record, in Rigby, J.K., and Hamblin, W.K., eds., Recognition of ancient sedimentary environments: Soc. Econ. Paleon. and Mineral., Spec. Publ. 16, p. 63-83.

Burggraf, D.R., Jr., 1976, Stratigraphy of the Upper Member, Koobi Fora Formation, southern Karari Escarpment, East Turkana Basin, Kenya (Master of Science Thesis): Iowa State University, Ames, Iowa.

Butzer, K.W., and Thurber, D.L., 1969, Some late Cenozoic sedimentary formations of the lower Omo basin: Nature, v. 222, p. 1138-1143.

Cerling, T.E., Biggs, D.L., Vondra, C.F., and Svec, H.L., 1975, Use of oxygen isotope ratios in correlation of tuffs, East Rudolf Basin, northern Kenya: Earth Planet. Sci. Lett., v. 25, p. 291-296.

Coleman, J.M., 1969, Brahmaputra River: Channel processes and sedimentation: Sed. Geol., v. 3, p. 131-239.

Coleman, J.M., and Gagliano, S.M., 1965, Sedimentary structures: Mississippi River deltaic plain, in Middleton, G.V., ed., Primary structures and their hydrodynamic interpretation: Soc. Econ. Paleon. Mineral., Spec. Publ. 12, p. 133-148.

Cooke, H.B.S., 1976, Suidae from Plio-Pleistocene strata of the Rudolf Basin, in Coppens, Y., Howell, F.C., Isaac, G.L., and Leakey, R.E.F., eds., 1976, Earliest man and environments in the Lake Rudolf Basin: University of Chicago Press, Chicago, p. 251-263.

Coppens, Y., Howell, F.C., Isaac, G.L., and Leakey, R.E.F., eds., 1976, Earliest man and environments in the Lake Rudolf Basin--Stratigraphy, paleoecology, and evolution: University of Chicago Press, Chicago, Illinois.

Curtis, G.H., Drake, T., Cerling, T.E., and Hampel, J.H., 1975, Age of KBS Tuff in Koobi Fora Formation, East Rudolf, Kenya: Nature, v. 258, p. 395-398.

Davies, D.K., Ethridge, F.G., and Berg, R.R., 1971, Recognition of barrier environments: Amer. Assoc. Petroleum Geologists Bull., v. 55, p. 550-565.

Deer, W.A., Howie, R.A., and Zussman, J., 1966, An introduction to the rock-forming minerals, Longman, London.

DeHeinzelin, J., Brown, F.H., and Howell, F.C., 1971, Plio-cene/Pleistocene formations in the lower Omo basin, southern Ethiopia: Quarternaria, v. 13, p. 247-268.

Dickinson, K.A., Berryhill, H.L., and Hohmes, C.W., 1972, Criteria for recognizing ancient barrier coast lines, in Rigby, J.K., and Hamblin, W.K., eds., Recognition of ancient sedimentary environments: Soc. Econ. Paleon. Mineral., Spec. Publ. 16, p. 192-214.

Dodson, R.G., 1963, Geology of the South Horr area: Geol. Surv. Kenya Report 60.

Findlater, I.C., 1976, Stratigraphic analysis and paleoenvironmental interpretation of a Plio/Pleistocene sedimentary basin east of Lake Turkana (Ph.D. dissertation): University of London, Birkbeck College, London.

_____, 1978, Stratigraphy, in Leakey, M.G., and Leakey, R.E.F., eds., Koobi Fora Research Project, Vol. 1, Clarendon Press, Oxford, p. 14-31.

Fitch, F.J., and Miller, J.A., 1976, Conventional potassium-argon and argon-40/argon-39 dating of volcanic rocks from East Rudolf, in Coppens, Y., Howell, F.C., Isaac, G.L., and Leakey, R.E.F., eds., Earliest man and environments in the Lake Rudolf Basin: University of Chicago Press, Chicago, p. 123-147.

Fitch, F.J., Hooker, P.J., and Miller, J.A., 1976, [40]Ar/[39]Ar dating of the KBS Tuff in the Koobi Fora Formation, East Rudolf, Kenya: Nature, v. 263, p. 740-744.

Folk, R.L., 1974, Petrology of sedimentary rocks: Hemphill's, Austin, Texas.

Frank, H.J., 1976, Stratigraphy of the Upper Member, Koobi Fora Formation, northern Karari Escarpment, East Turkana Basin, Kenya (Master of Science Thesis): Iowa State University, Ames, Iowa.

Fuchs, V.E., 1934, The geological work of the Cambridge expedition to the East African lakes: Geol. Mag., v. 71, p. 97-112 and p. 145-166.

Fuchs, V.E., 1935, The Lake Rudolf Rift Valley Expedition, 1934: Geograph. Jour., v. 86, p. 114-142.

Fuchs, V.E., 1939, The geological history of the Lake Rudolf basin, Kenya Colony: Phil. Trans. Roy. Soc. London, v. 229, p. 219-274.

Gautier, A., 1976, Assemblages of fossil freshwater mollusks from the Omo Group and related deposits in the Lake Rudolf Basin, in Coppens, Y., Howell, F.C., Isaac, G.L., and Leakey, R.E.F., eds., Earliest man and environments in the Lake Rudolf Basin: University of Chicago Press, Chicago, p. 379-382.

Harms, J.C., 1975, Stratification and sequence in prograding shoreline deposits, in Harms, J.C., Southard, J.B., Spearing, D.R., and Walker, R.G., eds., Depositional environments as interpreted from primary sedimentary structures and stratification sequences: Soc. Econ. Paleon. Mineral., Short Course 2, p. 81-102.

Harms, J.C., and Fahnestock, R.K., 1965, Stratification bed forms and flow phenomena, in Middleton, G.V., ed., Primary structures and their hydrodynamic interpretation: Soc. Econ. Paleon. Mineral., Spec. Publ. 12, p. 84-115.

Harris, J.M., 1976, Bovidae from the East Rudolf succession, in Coppens, Y., Howell, F.C., Isaac, G.L., and Leakey, R.E.F., eds., Earliest man and environments in the Lake Rudolf Basin: University of Chicago Press, Chicago, p. 293-301.

Harris, J.M., and White, T.D., 1977, Suid evolution and correlation of African hominid localities: Science, v. 198, p. 13-21.

Harrison, J.J., 1901, A journey from Zeila to Lake Rudolf: Geograph. Jour., v. 18, p. 258-275.

Hay, R.L., 1966, Zeolites and zeolitic reactions in sedimentary rocks: Geol. Soc. Amer. Spec. Paper 85.

_____, 1976, Geology of Olduvai Gorge, University of California Press, Berkeley.

Hillhouse, J.W., Ndombi, J.W.M., Cox, A., and Brock, A., 1977, Additional results on paleomagnetic stratigraphy of the Koobi Fora Formation, east of Lake Turkana (Lake Rudolf), Kenya: Nature, v. 265, p. 411-415.

Höhnel, L.V., Rosiwal, A., Toula, F., and Suess, E., 1891, Beitrage zur geologischen Kenntniss des ostilichen Afrika: Denkschriften der Kaiserlichen Akademie der Wissenschaften, v. 58, p. 447-584.

Höhnel, L.V., 1894, Discovery of Lakes Rudolf and Stephanie, Longmans, Green and Co., London.

Hurford, A.J., Gleadow, A.J.W., and Naeser, C.W., 1976, Fission-track dating of pumice from the KBS Tuff, East Rudolf, Kenya: Nature, v. 263, p. 738-740.

Johnson, G.D., 1974, Cainozoic lacustrine stromatolites from hominid-bearing sediments east of Lake Rudolf, Kenya: Nature, v. 247, p. 520-522.

Johnson, G.D., and Raynolds, R.G.H., 1976, Late Cenozoic environments of the Koobi Fora Formation: The Upper Member along the western Koobi Fora Ridge, in Coppens, Y., Howell, F.C., Isaac, G.L., and Leakey, R.E.F., eds., Earliest man and environments in the Lake Rudolf Basin: University of Chicago Press, Chicago, p. 115-122.

Joubert, P., 1966, Geology of the Loperot Area: Geol. Surv. Kenya Report 74.

Kanes, W.H., 1970, Facies and development of the Colorado River delta in Texas, in Morgan, J.P., ed., Deltaic sedimentation, modern and ancient: Soc. Econ. Paleon. Mineral., Spec. Publ. 15, p. 78-106.

Kendall, C.G., and Skipwith, P.A., 1968, Recent algal mats of a Persian Gulf lagoon: Jour. Sed. Petrol. v. 38, p. 1040-1058.

Leakey, R.E.F., Behrensmeyer, A.K., Fitch, F.J., Miller, J.A., and Leakey, M.D., 1970, New Hominid remains and early artifacts from northern Kenya: Nature, v. 226, p. 223-230.

LeBlanc, R.J., 1972, Geometry of sandstone reservoir bodies, in Cook, T.D., ed., Underground waste management and environmental implications: Amer. Assoc. Petroleum Geologists, Memoir 18, p. 133-190.

Leeder, M.R., 1978, A quantitative stratigraphic model for alluvium, with special reference to channel deposit density and interconnectedness, in Miall, A.D., ed., Fluvial sedimentology: Can. Soc. Petrol. Geol., Memoir 5, p. 587-596.

Leopold, L.B., and Wolman, M.G., 1957, River channel patterns; braided, meandering, and straight: U.S. Geol. Surv. Prof. Paper 282B.

Logan, B.W., Rezak, R., and Ginsburg, R.N., 1964, Classification and environmental significance of algal stromatolites: Jour. Geol., v. 72, p. 68-84.

Maglio, V.J., 1972, Vertebrate faunas and chronology of hominid-bearing sediments east of Lake Rudolf, Kenya: Nature, v. 239, p. 379-385.

Mathisen, M.E., 1977, A provenance and environmental analysis of the Plio-Pleistocene sediments in the East Turkana Basin, Lake Turkana, Kenya (Master of Science Thesis): Iowa State University, Ames, Iowa.

McConnell, R.B., 1972, Geological development of the rift system of eastern Africa: Geol. Soc. Amer. Bull., v. 83, p. 2549-2572.

McKenzie, D.P., Davies, D., and Molnar, P., 1970, Plate tectonics of the Red Sea and East Africa: Nature, v. 226, p. 243-248.

Patterson, B., Behrensmeyer, A.K., and Sill, W.D., 1970, Geology and fauna of a new Pliocene locality in northwestern Kenya: Nature, v. 226, p. 918-921.

Reineck, H.E., and Singh, I.B., 1975, Depositional sedimentary environments, Springer-Verlag, New York.

Schumm, S.A., 1968, Speculations concerning paleohydrologic controls of terrestrial sedimentation: Geol. Soc. Amer. Bull., v. 79, p. 1575-1588.

Simons, D.B., and Richardson, E.V., 1961, Forms of bed roughness in alluvial channels: Am. Soc. Civil Engineers, Proc., v. 87, p. 87-105.

Smith, A.D., 1896, Expedition through Somaliland to Lake Rudolf: Geograph. Jour., v. 8, p. 120-137 and p. 221-239.

Steel, R.J., 1974, New Red Sandstone flood plain and piedmont sedimentation in the Hebridean province, Scotland: Jour. Sed. Petrol., v. 44, p. 336-357.

Suess, E., 1891, Die Bruche des aestlichen Afrika: Denkschr. Akad. Wiss. Wien Math.-nat., v. 58, p. 555-584.

Vandamme, D., and Gautier, A., 1972, Molluscan assemblages from the Late Cenozoic of the lower Omo basin Ethiopia: Quarternary Research, v. 2, p. 25-37.

Vondra, C.F., and Bowen, B.E., 1976, Plio-Pleistocene deposits and environments, East Rudolf, Kenya, in Coppens, Y., Howell, F.C., Isaac, G.L., and Leakey, R.E.F., eds., Earliest man and environments in the Lake Rudolf Basin: University of Chicago Press, Chicago, p. 79-93.

Vondra, C.F., and Bowen, B.E., 1978, Stratigraphy, sedimentary facies, and paleoenvironments, East Rudolf, Kenya, in Bishop, W.W., ed., Background to early man in Africa: Geol. Soc. London, Spec. Publ. 5, p. 395-414.

Vondra, C.F., and Burggraf, D.R., Jr., 1978, Fluvial facies of the Plio-Pleistocene Koobi Fora Formation, Karari Ridge, East Lake Turkana, Kenya, in Miall, A.D., ed., Fluvial sedimentology: Can. Soc. Petrol. Geol., Memoir 5, p. 511-529.

Vondra, C.F., Burggraf, D.R., Jr., and White, H.J., 1978, The Plio-Pleistocene: sediments, environments, and geochronology along the Karari Escarpment, East Turkana, Kenya: Nebraska Acad. Sci., Trans., v. 6, p. 19-34.

Vondra, C.F., Johnson, G.D., Behrensmeyer, A.K., and Bowen, B.E., 1971, Preliminary stratigraphical studies of the East Rudolf basin, Kenya: Nature, v. 231, p. 245-248.

Walker, R.G., 1975, From sedimentary structures to facies models: Examples from fluvial environments, in Harms, J.C., Southard, J.B., Spearing, D.R., and Walker, R.G., eds., Depositional environments as interpreted from primary sedimentary structures and stratification sequences: Soc. Econ. Paleon. Mineral., Short Course 2, p. 63-80.

Walker, R.G., and Harms, J.C., 1971, The "Catskill Delta": A prograding muddy shoreline in central Pennsylvania: Jour. Geol., v. 79, p. 381-399.

Walsh, J., and Dodson, R.G., 1969, Geology of northern Turkana: Geol. Surv. Kenya Report 82.

White, J.J., 1976, Stratigraphy of the Lower Member, Koobi Fora Formation, southern Karari Escarpment, East Turkana Basin, Kenya (Master of Science Thesis): Iowa State University, Ames, Iowa.

Williams, P.F., and Rust, B.R., 1969,   The sedimentology of a braided river: Jour. Sed. Petrol., v. 39, p. 649-679.

Wunderlich, F., 1972,   Georgia coastal region, Sapelo Island, U.S.A. sedimentology and biology III, beach dynamics and beach development: Senckenbergiana Marit., v. 4, p. 47-79.

Daniel R. Burggraf, Jr., Howard J. White,
Hal J. Frank, Carl F. Vondra

4. Hominid Habitats in the Rift Valley: Part 2

Abstract

Lake Turkana, in the northwestern quarter of Kenya, is one of many African Lakes developed in depressions formed by crustal downwarping associated with Cenozoic continental rifting. The lake occupies a half-graben bounded on the west by a zone of major normal faulting and on the east by a westward dipping monoclinal flexure. During the Pliocene and Pleistocene epochs, the eastern shoreline of the paleo-Lake Turkana fluctuated in response to movement along the marginal faults which raised or lowered the half-graben floor. Consequently, more than 300 m of lacustrine, transitional, and fluvial sediments were deposited as a westwardly prograding sequence. These sediments record the gradual shift of the lake's eastern shoreline to the west. Volcanic and intrusive igneous highlands to the east and northeast provided large amounts of sediment which entered the basin in shallow, gravelly, ephemeral braided rivers. Volcaniclastic cobbles accumulated as channel bar and bar-core gravels which became sites for hominid tool-making activities. Aggradation of the basin and decreased tectonic activity resulted in the development of a broad, low relief flood plain, whose fine-grained sediments contain evidence of hominids in East Africa.

Introduction

As described in Part I, the East Turkana Basin consists of a thick (to 325 m) series of complexly intertongued Plio-Pleistocene sediments that record a general regression of paleo-Lake Turkana's eastern shoreline. The sedimentary blanket includes a thick interval of fine- and coarse-grained continental detritus which spans, albeit discontinuously, the time interval from greater than 3 million years ago to the

present.   Consequently, the deposits of the East Turkana
Basin enclose evidence of the activities of hominids as well
as their skeletal remains during a period critical to the
understanding of human evolutionary development.

The search for evidence of hominids in the East Turkana
Basin began in earnest when the National Museums of Kenya and
the Kenyan Government supported an expedition in 1968
(Leakey, 1970).   Since that time, scientists from many
countries have investigated the paleontological, anthropological, archaeological, and geological significance of the East
Turkana sediments, and although much insight has been gained
regarding the activities of hominids and their environments
in northern Kenya (Coppens et al., 1976; Leakey and Leakey,
1978), a multidisciplinary research effort continues in the
area under the auspices of the National Museums of Kenya and
the Kenyan Government.

Geological investigations were initiated in 1969 by
A. K. Behrensmeyer and expanded in 1970 to include the Iowa
State University Stratigraphic Research Group under the
direction of Dr. Carl F. Vondra.   Since 1975 the geological
program has been headed by Dr. Ian Findlater of the National
Museums of Kenya.   The program has included regional stratigraphic syntheses (Bowen and Vondra, 1973; Bowen, 1974;
Vondra and Bowen, 1978; Findlater, 1976, 1978); radiometric
dating frameworks (Fitch and Miller, 1970, 1976; Curtis et
al., 1975; Fitch et al., 1976; Hurford et al., 1976); geochemical and mineralogical characterization of basin sediments (Cerling, 1977, 1979; Cerling et al., 1977; Mathisen,
1977); and microstratigraphic studies (Bainbridge, 1976;
Burggraf, 1976; Frank, 1976; White, 1976; Vondra and
Burggraf, 1978) aimed at delineating depositional environments associated with the archaeologically and anthropologi-
cally most significant areas within the basin.   Through such
studies, and those from related disciplines, it has been
possible to develop a preliminary understanding of hominid
habitats in a rift-valley setting.

Location and Geology of the Karari Escarpment Area

While the entire East Turkana Basin is of interest for
the paleontological and geological information which it
records, the Karari Escarpment area in the northeastern portion of the basin is of particular significance because of
its relative abundance of archaeological and anthropological
material.   Exposure of the sediments is unusually complete
compared to many other parts of the basin, and access by
motorized vehicle is facilitated by a few dirt tracks along
the length of the escarpment.

Description

As shown in Figure 1, the Karari Escarpment (also called the Karari Ridge) is a relatively flat-topped cuesta which trends toward the southwest from the western margin of the volcanics comprising the Suregei Cuesta. The escarpment is upheld along much of its 20 m length by basalt-cobble conglomerates or by a bed of volcanic ash known as the Karari Tuff. The steep west-facing cliffs of the escarpment and the badlands area adjacent to them expose about 45 m of sandstones, siltstones, mudrocks, conglomerates, and tuffs important for the included fossil materials. The badlands occur in a narrow swath generally less than 200 m wide and are sparsely vegetated by various species of the desert scrub Commiphora-Acacia community (Ojany and Ogendo, 1973). It is the combination of resistant tuffs and carbonate-cemented coarse sandstones, underlain by poorly consolidated and unconsolidated sands and silts, which, during seasonal rainstorms develops steep-sided gullies eroding into the escarpment wall. In the absence of resistant beds, the relief along the escarpment is subdued, these areas being characterized by gentle slopes and slightly thicker vegetation. Drainage of the area is by westward-flowing sand-bed ephemeral streams of low sinuosity. The channels are commonly less than 30 m wide and 2 m deep and consist of loose sand and gravel derived from the sediments of the escarpment. Vegetated alluvial islands, up to several tens of meters long in length, often divide the channels. These channels occasionally include a zone of water-saturated permeable sands that serves as a watering hole for the local wildlife. The banks of the major channels are lined by a variety of vegetation ranging from short grasses and small thorny shrubs to large arborescent Acacia. Away from the channels, little plant life survives except for flat-topped thorny Acacia, short grasses, and Commiphora.

Geologically, the Karari Escarpment area is comprised of sediments of the Plio-Pleistocene Koobi Fora Formation which includes interbedded granule-to-boulder conglomerates, coarse- to fine-grained sandstones, siltstones and claystones, bioclastic carbonates and tuffs. It has been divided into the Lower and Upper Members (Vondra and Bowen, 1978), both of which have provided evidence of hominid.

Stratigraphy of the Koobi Fora Formation

A detailed description of the stratigraphy of the Koobi Fora Formation, particularly of the Lower Member, is presented in Part I of this chapter and, except for a brief review, will not be reproduced here. Details of the

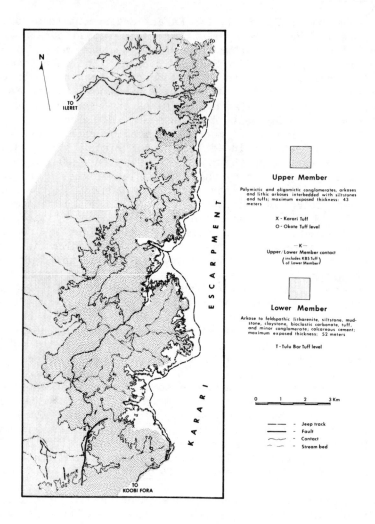

Figure 1.   Geology of the Karari Escarpment area.

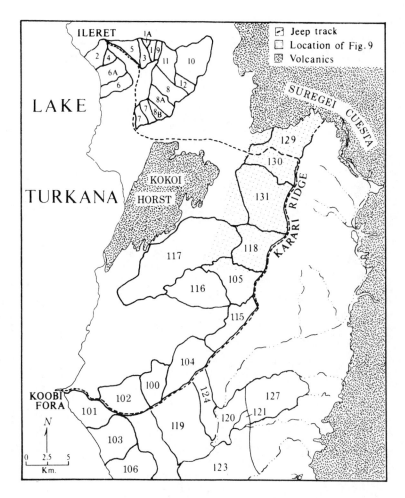

Figure 2.   Fossil collecting areas of the central East
Turkana Basin.   Reprinted by permission from
Vondra, Burggraf, and White (1978), Figure 6.

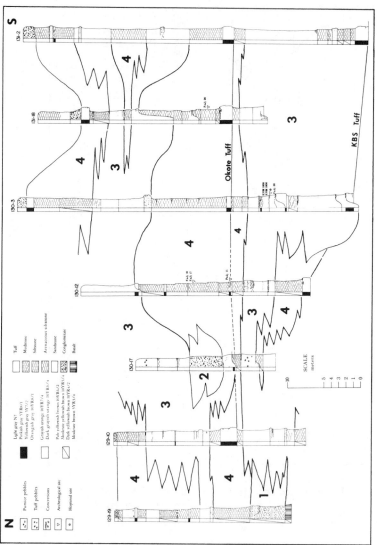

Figure 3. Graphic sections of the Upper Member, Koobi Fora Formation, northern Karari Escarpment. (1) Interbedded basalt conglomerate and pebbly mudstone subfacies; (2) Lenticular basalt clast conglomerate subfacies; (3) Polymictic conglomerate and sandstone subfacies; (4) Interbedded sandstone and tuffaceous siltstone subfacies

stratigraphy of the Upper Member will be presented as they
are of prime importance to the understanding and interpretation of the depositional environments extant in the East
Turkana Basin (Fig. 2) during a portion of the Pleistocene
epoch.

Lower Member. The Lower Member of the Koobi Fora
Formation along the Karari Escarpment reaches a maximum exposed thickness of about 135 m (Bowen, 1974), the upper half
of which has been studied in detail by White (1976) along the
southern portion of the Escarpment and by Bainbridge (1976)
to the north. Mudrock is the dominant lithology of the
member (White, 1976) which exhibits an indistinct overall
upward coarsening in average grain size. The Lower Member is
comprised of a complexly intertongued sequence of tabular
mudstones and siltstones, thin bioclastic carbonates and
intraformational limonite-clast conglomerates, and
lenticular-bedded mudstones and fine-grained sandstones
(White, 1976). In addition, the Lower Member sediments
include a major tuff horizon, the KBS, which occurs close to
the Lower Member/Upper Member boundary. This contact between
the two members is defined as the base of the conglomeratic
channel sandstone complex which occurs stratigraphically
immediately above the KBS Tuff (Vondra and Bowen, 1978).

Upper Member. Along the Karari Escarpment, the Upper
Member disconformably overlies the Lower Member and is
characterized by large-scale trough and planar cross-
stratified lithic arkoses, and feldspathic litharenites,
basalt pebble-to-cobble conglomerates, tuffaceous siltstones,
tuffs, and mudstones (Burggraf, 1976; Frank, 1976). The
distinct channeling of the Upper Member into the Lower Member
is marked by as much as 8 m of relief along the northern portion of the Karari Escarpment (Frank, 1976) and by a maximum
of about 11 m in area 105 (Burggraf, 1976). From area 129
the Upper Member thickens southward to a maximum of about 45
m in area 131 (Frank, 1976) and then thins to less than 22 m
in northern area 105. Lateral relationships are illustrated
in Figures 3 and 4.

The lower half of the Upper Member consists almost
exclusively of poorly consolidated sandstones and lenses of
basalt granule-to-cobble conglomerates. Individual conglomerate beds can be thicker than 4 m, occurring as massive or
indistinctly bedded units, but more commonly are less than 2
m thick. They also display indistinct imbrication and large-
scale planar cross-stratification and interfinger with
conglomeratic sandstones and coarse-grained sands. Matrix-
supported conglomerates occur as 0.4 m thick units with
indistinct upper and lower contacts in and northeast of area

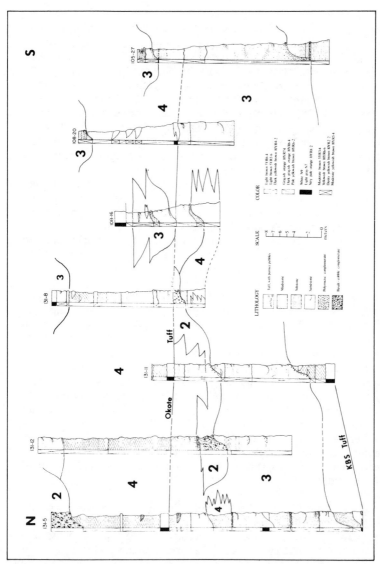

Figure 4. Graphic sections of the Upper Member, Koobi Fora Formation, southern Karari Escarpment. (1) Interbedded basalt conglomerate and pebbly mudstone subfacies; (2) Lenticular basalt clast conglomerate subfacies; (3) Polymictic conglomerate and sandstone subfacies; (4) Interbedded sandstone and tuffaceous siltstone subfacies

129. Within the sandstones, internal scour surfaces are common, and lateral variation in primary sedimentary structures is rapid. These include well-developed large-scale planar and trough cross-stratification, small-scale trough cross-stratification, plane bedding, and ripple-lamination. From north to south, the dominant sandstone composition grades from feldspathic litharenite through lithic arkose to arkose (Burggraf, 1976; Frank, 1976). Calcrete horizons occur throughout this coarse-grained sequence but are best developed in the finer sands which often cap a fining-upward sand sequence. Carbonate-cemented crotovina and/or root casts frequently occur with the calcretes and may be as much as 3 or 4 cm thick and 50 cm long but most commonly are on the order of 1 cm in diameter and 10 to 20 cm in length.

The upper half of the Upper Member consists of tuffaceous siltstones, mudstones, and fine-grained sandstones which occur in beds distinctly more tabular than those of the underlying sands and conglomerates. The basal contact of this fine-grained interval is apparently conformable and is marked either by the Okote Tuff or by a gradation from poorly sorted sand to tuffaceous siltstone. As in the underlying sediments, the beds of the upper half of the Upper Member contain abundant primary sedimentary structures ranging from large-scale trough cross-stratification to ripple-lamination and plane bedding. Mudstones often contain abundant irregular carbonate concretions a few centimeters in diameter, and siltstones and sandstones include carbonate-cemented crotovina and root casts (Burggraf, 1976; Frank, 1976).

Tuffs found interbedded with the sediments of the Upper Member are important because they are believed to represent a relatively short period of time in the depositional history of the basin. As such, they are interpreted to be isochronous units (Findlater, 1976, 1978) which, when containing minerals with sufficiently high quantities of certain radioisotopic elements (such as potassium), can be dated for the purpose of establishing an absolute time framework. Tuffs within the Upper Member of the Koobi Fora Formation often contain pummice lumps which include crystals of feldspar suitable for dating. The best dates for the Okote and Karari tuffs are, respectively, 1.56 $\pm$ 0.02 m.y. and 1.32 $\pm$ 0.10 m.y. (Fitch and Miller, 1976). While the ages of some of the other tuff horizons within the Koobi Fora Formation are controversial, the above ages for the Okote and Karari tuffs are widely accepted.

Facies and Subfacies of the Koobi Fora Formation

The strata surrounding the Karari Escarpment belong to

Subfacies of the Lenticular Conglomerate, Sandstone and
Mudstone Facies:

1) interbedded basalt conglomerate   (*Alluvial fan; debris flow*)
   and pebbly mudstone

2) lenticular basalt clast conglomerate

                                                    (*Channel*)

3) polymictic conglomerate and sandstone

4) interbedded sandstone and tuffaceous siltstone   (*Floodplain*)

Figure 5.   Subfacies of the lenticular conglomerate, sandstone,
            and mudstone facies. Reprinted by permission from
            Vondra and Burggraf (1978), Figure 4.

four major facies of the Koobi Fora Formation. These are
the: 1) laminated siltstone facies, 2) arenaceous bioclastic
carbonate facies, 3) lenticular fine-grained sandstone and
lenticular bedded siltstone facies, and 4) lenticular conglomerate sandstone and mudstone facies (Bowen and Vondra,
1973; Vondra and Bowen, 1976, 1978; and Part I, this volume).
The sediments comprising the Upper Member are a part of the
lenticular conglomerate, sandstone, and mudstone subfacies
which represents deposition in fluvial channel and flood
basin environments (Vondra and Burggraf, 1978). Within this
subfacies, four fluvial subfacies have been described (Vondra
and Burggraf, 1978). These are the: 1) interbedded basalt
cobble conglomerate and pebbly mudstone subfacies, 2) lenticular basalt clast conglomerate subfacies, 3) polymictic
conglomerate and sandstone subfacies, and 4) interbedded
sandstone and tuffaceous siltstone subfacies (Fig. 5).
Recognition of these subfacies is based on field mapping and
measurement of bed geometries, primary sedimentary structures, and areal and vertical distribution of the major
lithotypes. The four subfacies are complexly related both in
time and space but, in general, occur as a progressive
sequence representing basin-margin to basin-center depositional environments.

Interbedded Basalt Cobble Conglomerate and Pebbly Mudstone Subfacies

The interbedded basalt cobble conglomerate and pebbly
mudstone subfacies occur along the basin margin in the
northern portion of area 129 thinning to the southwest from a
thickness in excess of 15 m (Vondra and Burggraf, 1978). Bed
thicknesses vary from 0.4 m clast and/or matrix-supported
units having indistinct upper and lower bedding planes to 2.0
m clast-supported lenses which are several tens of meters
long and exhibit distinct erosional surfaces (Frank, 1976).
The subrounded-to-rounded pebbles and cobbles of basalt and
ignimbrite occur in a matrix of arenaceous mudstone or
coarse-grained litharenite and may be arranged as crude horizontal stratification or indistinct large-scale planar cross
stratification within the clast-supported conglomerates.
Matrix-supported units are arenaceous mudstones with abundant
volcanic rock fragments, isolated pebbles of basalt and
ignimbrite, and limonitic concretions (Frank, 1976).

The interbedded basalt cobble conglomerate and pebbly
mudstone subfacies comprise a wedge-shaped sequence of
detritus which thins to the south and west interfingering
with, and grading into, conglomeratic sandstone and lenticular basalt clast conglomerate. The location of this subfacies adjacent to the volcanic source area directly to the

east; the coarseness of the sediments; the occurrence of matrix-supported conglomerates interbedded with clast-supported units, and the westward intertonguing with conglomerates and sandstones toward the basin center suggest that sediments of the interbedded basalt cobble conglomerate and pebbly mudstone subfacies represent an ancient alluvial fan system similar to those described by Blissenbach (1954), Hooke (1967), Bull (1964a, 1964b, 1968, 1972), and Walker (1975). Massive or crudely bedded gravels (facies "Gm" of Miall, 1977) are interpreted to represent gravel sheets deposited by sheetflooding across the surface of the alluvial fan during periods of high rainfall. Planar cross-stratified gravels (facies "Gp" of Miall, 1977) with lenticular geometries developed in ephemeral stream channels following flood conditions (Bull, 1964a, 1968, 1972). Pebbly mudstones (facies "Gms" of Miall, 1978) and arenaceous mudstones of this subfacies are interpreted to represent debris flow deposits indicative of low annual precipitation but with irregular and brief intervals of heavy rainfall onto a sparsely vegetated slope analogous to coarse-grained alluvial fan sedimentation described by Bull (1972).

Lenticular Basalt Clast Conglomerate Subfacies

The lenticular basalt clast conglomerate subfacies consists of dominantly structureless, loose to poorly indurated, laterally discontinuous lenticular beds up to 4.3 m thick (Vondra and Burggraf, 1978). The subfacies grades basin-ward from clast- and matrix-supported conglomerates of the interbedded basalt cobble conglomerate and pebbly mudstone subfacies. The pebbles and cobbles are well rounded and subspherical (Fig. 6) and occur within a matrix of coarse sand consisting of subangular clasts of quartz, feldspar, gneiss, granite, and ignimbrite (Burggraf, 1976). Infrequently included with these are clasts of rounded carbonate-cemented sandstone or basalt pebbles, which suggest reworking of older calcrete horizons. Exposures of this subfacies often include dense carbonate-cemented beds of conglomeratic sandstone to 0.2 m thick and less densely cemented beds less than 0.5 m thick (Vondra and Burggraf, 1978). Primary sedimentary structures are usually limited to indistinct horizontal stratification and normally graded bedding, but large-scale planar cross-stratification is found as high-angle sets up to 0.5 m thick, commonly overlain by a thinner lower angle set. Interfingering and interbedded wedges of coarse-grained sands occur in most exposures of this subfacies and exhibit small-scale trough cross-stratification, carbonate-cemented root casts and crotovina, and frequently include abraded mammalian skeletal remains.

Figure 6. Photographs of the fluvial subfacies of the Upper
Member. (a) Outcrop of the lenticular basalt clast
conglomerate subfacies; (b) Large scale planar
cross strata of the polymictic conglomerate and
sandstone subfacies; (c) Large scale trough cross
strata of the polymictic conglomerate and sandstone subfacies

Figure 6, continued. (d) Caliche horizon with carbonate-
            cemented root casts extending into underlying
            silty sand of the polymictic conglomerate and
            sandstone subfacies. (e) Upper/Lower Member
            contact in area 131.

Figure 6, continued. (f) Calcareous root casts of the interbedded sandstone and tuffaceous siltstone subfacies
(g) Outcrop of the Karari Tuff in area 131; note
laminations

The deposits of the lenticular basalt clast conglomerate subfacies include the fluvial facies "Gm," "Gp," and "St" of Miall (1977) and are interpreted to represent major channels of dominantly braided streams which drained the alluvial fans and adjacent volcanic highlands. Deposits having similar features have been described by Eynon and Walker (1974), Rust (1975), Boothroyd and Ashley (1975), and Boothroyd and Nummedal (1978). Horizontally stratified gravels may represent the "bar cores" of Eynon and Walker (1974), and the high angle planar cross-stratified gravels overlain by lower angle planar cross-strata are interpreted to represent transverse bars in a gravelly braided river. The apparent lack of very large-scale bedforms similar to those described by Eynon and Walker (1974) indicates that the channels of the lenticular basalt clast conglomerate subfacies were dominantly shallow, probably less than 1 m deep, and that the discharge was low compared to the modern gravelly, braided outwash rivers described by Rust (1975), Boothroyd and Ashley (1975), and Boothroyd and Nummedal (1978).

Polymictic Conglomerate and Sandstone Subfacies

The thick sequence of lithic arkoses, arkoses, and feldspathic lithaenites comprising the lower portion of the Upper Member consists of large-scale planar and trough cross-stratified coarse- to fine-grained sands, horizontally stratified medium to coarse-grained sands, and ripple-laminated to festoon cross-stratified fine-grained sand (Fig. 6). Individual beds average about 1.5 m in thickness, have erosional bases, and are normally graded. Three dominant primary structure groups have been described (Vondra and Burggraf, 1978). These include solitary sets and cosets of large-scale trough cross-stratified sand to 1 m thick; tabular to wedge-shaped large-scale planar cross-stratified sand in sets from 5 to 30 cm thick; and a sequence consisting of a basal unit of structureless conglomeratic sand grading upward into large-scale trough cross-stratified sand, and horizontally stratified or small-scale trough cross-stratified sand and silt, commonly overlain by a veneer of mudrock. A complete sequence of this last group may reach a thickness of nearly 5 m, but often the upper portions are not preserved, due to erosion by later events. Other structures preserved in the polymictic conglomerate and sandstone subfacies include carbonate-cemented root casts and crotovina, clay balls, and rip-up clasts in the basal part of coarse-grained sandstone units and abraded vertebrate skeletal remains.

The deposits of the polymictic conglomerate and sandstone subfacies are interpreted as representing large-scale dune and transverse, longitudinal and point-bar environments of deposition (Vondra and Burggraf, 1978). The first two primary sedimentary structure groups described above are believed to represent large-scale bedforms in the channels of coarse-grained, sandy braided rivers similar to those described by Ore (1963), Allen (1965b, 1970), Williams and Rust (1969), Williams (1971), Kessler (1971), Rust (1972), Smith (1970, 1971, 1974), and consist of the general fluvial facies "St," "Sp," "Sr," "Sh," and "Ss" of Miall (1977). These include scour-and-fill features and longitudinal and transverse bars composed of horizontally stratified to ripple-laminated sand and large-scale trough and planar cross-stratified sands (Vondra and Burggraf, 1978). The third primary structure group is most common in the strati-graphically higher portions of the Upper Member and repre-sents point bar deposition in a bed load, sandy, interme-diate- to high-sinuosity river. Such a point bar sequence was first described by Bernard and Major (1963) and later formalized by Allen (1970). The occurrence of this sequence toward the top of this subfacies indicates a change in the fluvial regime to lower energy meandering conditions.

The polymictic conglomerate and sandstone subfacies interfingers with the lenticular basalt clast conglomerate subfacies and is overlain by the interbedded sandstone and tuffaceous siltstone subfacies. Along the Karari Escarpment the polymictic conglomerate and sandstone subfacies reaches a maximum thickness of about 20 m.

Interbedded Sandstone and Tuffaceous Siltstone Subfacies

The upper half of the Upper Member is composed of the interbedded sandstone and tuffaceous siltstone subfacies which includes tabular beds of tuffaceous siltstone, fine-to medium-grained sandstones, claystones, and tuffs. These sediments occur as a series of fining-upward cycles, con-sisting of up to 2 m of massive to horizontally stratified or large-scale trough cross-stratified, fine- to medium-grained sandstone grading upward into several meters of massive, parallel-laminated and ripple-laminated, arenaceous and tuf-faceous siltstone. The siltstone includes abundant small calcareous root casts (less than 1 cm thick and several cen-timeters long), nodular carbonate concretions, and occa-sionally thin tuff lenses up to 0.5 m thick. The siltstone often grades upward into a silty mudstone up to 3 m thick, which is usually mottled and contains nodular carbonate con-cretions (Fig. 6) 1 to 4 cm in diameter (Vondra and Burggraf, 1978).

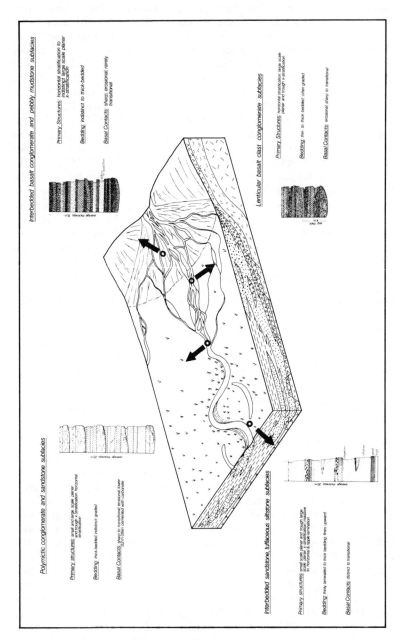

Interbedded basalt conglomerate and pebbly mudstone subfacies

Primary Structures: horizontal stratification to indistinct large scale planar x-stratification

Bedding indistinct to thick-bedded

Basal Contacts: sharp, erosional; rarely transitional

Lenticular basalt clast conglomerate subfacies

Primary Structures: horizontal stratification; large scale planar and trough x-stratification

Bedding: thin- to thick-bedded; often graded

Basal Contacts: erosional; sharp to transitional

Polymictic conglomerate and sandstone subfacies

Primary structures: small and large scale planar and trough x-stratification; horizontal stratification

Bedding: thick-bedded; indistinct; graded

Basal Contacts: sharp to transitional; erosional; lower 0.2m often cemented with carbonate

Interbedded sandstone, tuffaceous siltstone subfacies

Primary structures: small scale planar and trough, large scale planar x-stratification; massive to horizontal & ripple lamination

Bedding: thinly laminated to thick bedding; fines upward

Basal Contacts: distinct to transitional

Figure 7. Fluvial subenvironments of the Upper Member along the Karari Escarpment.

The lower and upper boundaries of the interbedded sandstone and tuffaceous siltstone subfacies are frequently marked by two prominent tuff horizons, the lower being the Okote Tuff and the upper the Karari Tuff. These consist primarily of light gray ash with variable amounts of included clastic debris and may be thicker than 2 m. The tuffs often include a lower massive unit about 0.5 m thick overlain by a parallel-laminated to small-scale planar or trough cross-stratified ash up to 1 m thick. The upper portion of the tuffs may include pumice pebbles enclosed in a ripple-laminated to massive ash usually less than 20 m thick (Vondra and Burggraf, 1978).

The fine-grained and tabular nature of the strata of the interbedded sandstone and tuffaceous siltstone subfacies and the interfingering relationship with coarser grained exposures of the polymictic conglomerate and sandstone subfacies suggests that deposition occurred on a low relief flood plain by a combination of vertical accretion, crevasse-splay flooding, and eolian reworking (Vondra and Burggraf, 1978). The features of ancient and modern flood plain deposits have been discussed by many authors including Wolman and Leopold (1957), Allen (1964, 1965a, 1965b, 1970), Leopold et al. (1964), Anderson and Picard (1974), and Steel (1974).

Depositional Environment and Paleogeography

The recognition of fluvial depositional subenvironments as delineated by subfacies reconstruction coupled with their areal distribution and intertonguing relationships produces a record of the paleogeographic development of the north central East Turkana Basin (Fig. 7).

The sediments of the Lower Member of the Koobi Fora Formation indicate dominant lacustrine, beach, and deltaic conditions which persisted in the East Turkana Basin for much of the period from about 3 m.y. to about 2 m.y. B.P. (see Part I, this volume). Geochemical, mineralogical investigations (Cerling, 1977, 1979) and paleontological studies (Williamson, 1978) have shown that throughout the deposition of the Lower Member, paleo-Lake Turkana was a fresh to slightly brakish water body with a diverse molluscan population. Between 1.5 and 1.8 m.y. B.P., however, a dramatic change began which initiated widespread deposition of fluvial sediments in the eastern portion of the basin. The paleo-Lake Turkana fluctuated between fresh and brakish conditions (Cerling, 1979) and endemic mollusc species became extinct (Williamson, 1978).

Uplift along the basin margins created highlands which

supplied large amounts of coarse clastic debris to the rivers which flowed into the lake from the northeast. Adjacent to the volcanic highlands bordering the basin on the east and northeast, broad alluvial fans developed as wedge-shaped deposits of the interbedded basalt cobble conglomerate and pebbly mudstone subfacies (Vondra and Burggraf, 1978). At least one major southwestwardly flowing stream entered the basin from the northeast carrying with it clastic debris from plutonic igneous and metamorphic terranes north and east of the volcanic highlands (Mathisen, 1977). This river system eroded channels as much as 11 m deep into the deltaic and flood-plain sands, silts, and clays of the upper portion of the Lower Member (Burggraf, 1976; White, 1976; Frank, 1976).

Rapid erosion of the highland areas supplied enough sediments, and, presumably, discharge was variable enough that a dominantly braided shallow river system developed in the north central portion of the East Turkana Basin. Low sinuosity channels migrated across the aggrading plain depositing coarse sands and gravels as longitudinal and transverse bars and gravel sheets (Vondra and Burggraf, 1978). These developed the lenticular basalt clast conglomerate and the polymictic conglomerate and sandstone subfacies. Deposition proceeded largely from sand-bed ephemeral streams as indicated by the numerous caliche horizons and calcified root casts and crotovina which developed in the channel sands during periods of low rainfall. At least one perennial drainage existed along the southern end of the present Karari Escarpment (Greenwood, personal communication regarding fish fauna, in Mathisen, 1977). The fine-grained component of these sediments is poorly recorded today but is found as rip-up clasts and clay balls in the lower portions of coarse-grained sand sequences.

Almost as abruptly as it began, the deposition of coarse sands and gravels ceased along the present day Karari Escarpment. The transition is marked by the Okote Tuff (about 1.5 m.y. B.P. ago) and by widespread occurrence of tuffaceous siltstones and fine-grained sandstones which are in conformable contact with the underlying sandstones and conglomerates. The sudden predominance of fine-grained sediments, and particularly of volcanic ash, suggests a lowering of the source area due to active tectonism. The decrease in relief between the basin and the surrounding highlands led to the development of a lower energy inter-mediate- to high-sinuosity river system as reflected by the series of fining-upward point bar sequences of the uppermost polymictic conglomerate and sandstone subfacies and the interbedded sandstone and tuffaceous siltstone subfacies. Vertical accretion of flood-plain silts and clays produced the tabular, massive- to ripple-laminated siltstones and clay-stones of the upper half of the Upper Member. Ubiquitous calcareous root casts and carbonate nodules in the finest sediments as well as loosely cemented carbonate horizons resembling caliche profiles indicate that during the deposition of the uppermost part of the Upper Member the climate of northern Kenya was arid to sub-arid (Frank, 1976; Burggraf, 1976). This is also supported by the authigenic mineral content of the Upper Member sediments and by the increased salinity of the lake chemistry (Cerling, 1977, 1979). Analysis of blocky, bioturbated, and mottled mudrocks often capping thick flood-plain siltstones (Fig. 8) suggests incipient soil development (White, 1976).

Aggradation of the basin was interrupted following deposition of the Karari Tuff. Toward the basin center, at least two subsequent episodes of lacustrine transgression occurred as preserved in the sediments of the Guomde Formation and the Galana Boi Beds (Bowen and Vondra, 1973). Along the Karari Escarpment, a short interval of Late Pleistocene fluvial deposition is preserved as a thin sequence of sandstones and conglomerates. In many places, this deposition has truncated the Karari Tuff and undergone extensive erosion since the last phase of regional uplift during the latest Pleistocene (Vondra and Burggraf, 1978).

The area today is drained by west and southwestwardly flowing, shallow, low-sinuosity, ephemeral streams. The sediment load consists primarily of silt, sand, and gravel derived locally from exposures of the Koobi Fora Formation.

Archaeological Sites and Hominid Habitats

As mentioned previously, many archaeological and hominid sites occur along the Karari Escarpment in sediments of the Upper Member of the Koobi Fora Formation (Fig. 9). Many of these have been studied in detail (Isaac, Leakey, and Behrensmeyer, 1971; Leakey and Harris, 1975; Isaac, 1976; Isaac, Harris, and Crader, 1976; Harris and Isaac, 1976; Harris and Herbich, 1978) and have been shown to have occurred in fluvial subenvironments (Frank, 1976; Findlater, 1976, 1978). In addition, several hominid sites have been discovered in the delta plain sediments of the Lower Member. Geological studies of these sites have been completed by Bainbridge (1976).

Two of the four fluvial subfacies are of particular importance for their archaeological and hominid material. The polymictic conglomerate and sandstone subfacies and the interbedded sandstone and tuffaceous siltstone subfacies

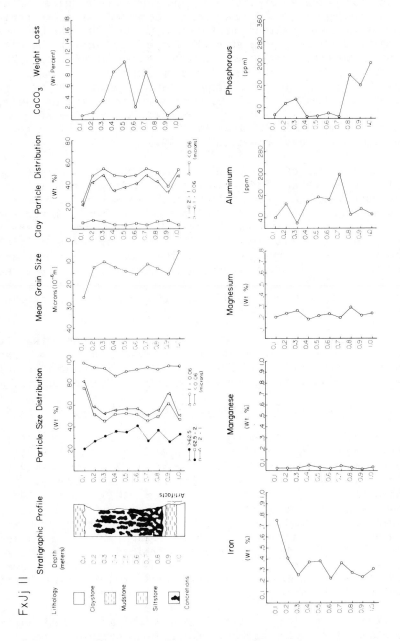

Figure 8. Analyses of a paleosolic horizon at FxJj 11.

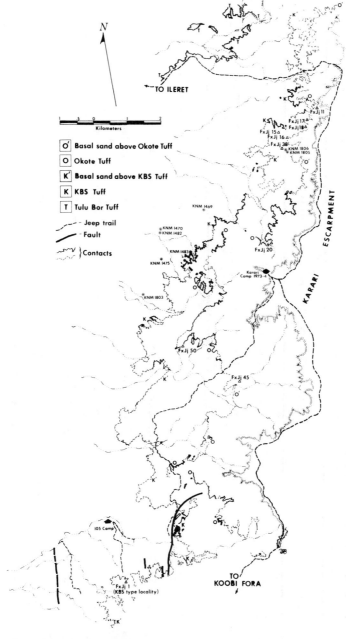

Figure 9. Archaeological and hominid sites along the Karari Escarpment. Reprinted by permission from Vondra, Burggraf, and White (1978), Figure 9.

Table 1. Archaeological sites, lithologic association, and depositional environment.

Site	Lithologic association	Environmental interpretation
FxJj 11	Tuffaceous mudstone	Flood plain (distal)
FxJj 17	Arenaceous siltstone	Flood plain (proximal)
FxJj 18 NS	Feldspathic litharenite	Channel (bar)
FxJj 18 IHS	Silty Tuff	Flood plain (proximal)
FxJj 18 GS	Feldspathic litharenite	Channel (bar)
FxJj 16	Lithic arkose	Channel (thalweg)
FxJj 15	Feldspathic litharenite	Channel (thalweg)
FxJj 38	Lithic arkose	Channel
FxJj 20 AB, M, and E	Arenaceous siltstone	Flood plain
FxJj 23	Lithic arkose	Channel

enclose artifacts, creating sparse associations of a few
fragments or close groups of up to several hundred pieces
(Frank, 1976). Table 1 lists a number of individual archaeological sites with the dominant associated lithologies and
environmental interpretations.

Commonly, the artifact horizons are found in the lower
portions of fining-upward sand and silt sequences (FxJj 15;
FxJj 18 N.S.) or at the contact between a sandstone and a
tuffaceous mudstone (FxJj 11; FxJj 17). The surrounding
strata are often mottled and may contain abundant columnar
carbonate root casts and/or crotovina which may grade
laterally into more densely cemented calcrete horizons. These
features suggest an incipient soil profile (Fig. 7, FxJj 11)
which developed on a flood-plain site adjacent to an
ephemeral stream channel. Other sites are in direct association with lenses of basalt cobbles and conglomeratic coarse-
grained sandstone sequences (FxJj 18 G.S.; FxJj 16). These
are interpreted to represent gravel bars in a coarse-grained
braided river which served as tool-making sites for the hominids (Harris and Herbich, 1978).

From these and other studies, it is clear that evidence
of hominids in northern Kenya is preserved in a variety of
sediments and depositional environments ranging from the
gravels of alluvial fans to the sands and silts of lacustrine
beaches and delta plains. Harris and Herbich (1978) studied
the distribution of more than 50 archaeological sites along
the Karari and Ileret ridges and have associated hominid
activities with:

1) "low-lying channel and flood-plain localities
   close to the lake margin;

2) stream channel and flood-plain localities on
   the alluvial plain, some 15 to 25 kilometers
   inland from the shoreline; and

3) alluvial-fan localities at the foot of the low
   hills at the rim of the East Lake Turkana Basin
   ..." (Harris and Herbich, 1978, p. 539).

It was further concluded that the distribution of sites is
related to the "...proximity of stream and river channels and
the accessibility of a source of raw material..." (Harris and
Herbich, 1978, p. 539) for tool-making activities. Additionally, the occupation of some sites may have been related to
seasonally high stream levels, and there appears to be
variability in stone-tool assemblages and preserved faunal

remains between channel sites and flood-plain sites (Harris and Herbich, 1978).

The features of archaeological and hominid sites noted by Harris and Herbich (1978) reflect aspects of hominid behavior and artifact preservability as a direct consequence of the rift-valley setting in which the hominids lived. As a rapidly aggrading sedimentary basin in an arid-to-semiarid climate, the East Turkana area received abundant volcaniclastic and plutonic igneous detritus from erosion of the uplifted marginal highlands to the north and east. Basalt and ignimbrite cobbles throughout alluvial-fan sediments and in coarse-grained channel bars provided a convenient and abundant supply of tool-making materials within the seasonally flooded water courses which sustained both plant and animal life in the basin. Preservation of the activities of the hominids is due to rapid burial of occupation sites and includes artifact and bone fragments in sands and silts from overbank floods. The apparent lack of faunal remains associated with channel sites as contrasted to that of flood basin silts and clays may be in part attributed to the higher energies associated with the former. Bed-load sands and gravels would certainly abrade and probably destroy most bone fragments associated with sites proximal to an active channel, whereas suspension sedimentation of overbank fines would bury and preserve, in a relatively undisturbed state, both stone artifacts and bone fragments abandoned in a distal flood-plain occupation site.

Episodic volcanism adjacent to the basin served not only to supply an abundance of tool-making material but also to provide large amounts of fine-grained sediments which aided in the burial and preservation of evidence of hominid activities. Abundant dissolved solids from devitrifying ash particles and from the vast basaltic highlands to the east likely slowed the deterioration of vertebrate faunal remains, as did the arid-to-semiarid climate which prevailed in the basin during much of the Pleistocene Epoch.

The relative abundance of hominid remains associated with the transitional lacustrine (arenaceous bioclastic carbonate facies; see Part I, this volume) and deltaic (lenticular fine-grained sandstone and lenticular-bedded siltstone facies; see Part I, this volume) sediments may be related to rapid sedimentation rates in an environment more suited to long-term occupation by hominids. The higher density and varied population of fauna which is found along the present shoreline is analogous to that which existed during much of the Pleistocene Epoch (with certain exceptions) but likely served as an important food source for the hominids. The relative lack of tool sites found associated with beach sediments is likely due to the absence of suitable materials close by. Also, any tool sites in this high energy regime would be dispersed.

Summary

The rift-valley tectonic setting as exemplified by the East Turkana Basin provided many requirements for the support and preservation of hominid activity in East Africa. Local crustal warping and volcanism within the rift-created depressions or blocked drainage, thus creating large lakes. The rift, with its streams and lakes, supported large populations of both plants and animals and served as a migratory route for fauna throughout much of the Late Cenozoic. During the Pleistocene Epoch, uplift of the margins of the East Turkana Basin provided vast amounts of volcaniclastic and plutonic igneous detritus which poured into the basin by means of a southwestwardly prograding braided stream system. Basalt and ignimbrite cobbles occurring in alluvial-fan gravels and in coarse-grained bars of braided rivers provided tool-making materials for the hominids. Rapid basin aggradation, transition from higher energy braided rivers to lower energy meandering rivers, and an abundant supply of fine-grained sediments, both from weathering of the igneous highlands to the east and northeast of the basin and from copious amounts of volcanic ash, provided a favorable environment for the burial and preservation of evidence of hominid activities in the East Turkana Basin.

Acknowledgments

Sincere gratitude is extended to Richard Leakey and the National Museums of Kenya who supported the field research through logistical aid and encouragement. This research was funded by National Science Foundation grants GA-25684, GS-37814, and EAR-7902800.

References

Allen, J.R.L., 1964, Studies in fluviatile sedimentation; Six cyclothemes from the lower Old Red Sandstone, Anglo-Welsh Basin: Sedimentology, v. 3, p. 163-198.

_____, 1965a, Fining-upwards cycles in alluvial successions: Geol. J., v. 4, p. 229-246.

_____, 1965b, A review of the origin and characteristics of recent alluvial sediments: Sedimentology, v. 5, p. 89-191.

_____, 1970, Studies in fluviatile sedimentation; a comparison of fining-upwards cyclothemes, with special reference to coarse-member composition and interpretation: J. Sediment. Petrol., v. 40, p. 298-323.

Anderson, D.W., and Picard, M.D., 1974, Evolution of synorgenic clastic deposits in the intermontane Uinta Basin of Utah, in Dickenson, W.R., ed., Tectonics and Sedimentation: Soc. Econ. Paleont. Mineral. Spec. Pub. 22, p. 167-189.

Bainbridge, R.B., 1976, Stratigraphy of the Lower Member, Koobi Fora Formation, northern Karari Escarpment, East Turkana Basin, Kenya (Master of Science Thesis): Iowa State University, Ames, Iowa.

Bernard, H.A., and Major, C.F., Jr., 1963, Recent meander belt deposits of the Brazos River; an alluvial "sand" model abs.: Am. Assoc. Petr. Geol. Bull., v. 47, p. 350-351.

Blissenbach, E., 1954, Geology of alluvial fans in semiarid regions: Geol. Soc. Am. Bull., v. 65, p. 175-190.

Boothroyd, J.C., and Ashley, G.M., 1975, Processes, bar morphology, and sedimentary structures on braided outwash fans, northeastern gulf of Alaska, in Jopling, A.V., and McDonald, B.C., eds., Glaciofluvial and Glaciolacustrine sedimentation: Soc. Econ. Paleont. Mineral. Spec. Pub. 23, p. 193-222.

_____, and Nummedal, D., 1978, Proglacial braided outwash; a model for humid alluvial fan deposits, in Miall, A.D., ed., Fluvial Sedimentology: Canadian Soc. Petr. Geol. Mem. 5, p. 641-668.

Bowen, B.E., 1974, The geology of the Upper Cenozoic Sediments in the East Rudolf embayment of the Lake Rudolf basin, Kenya (Ph.D. dissertation): Iowa State University, Ames, Iowa.

_____, and Vondra, C.F., 1973, Stratigraphical relationships of Plio-Pleistocene deposits, East Rudolf, Kenya: Nature, v. 242, p. 391-393.

Bull, W.B., 1964a, Alluvial fans and near-surface subsidence in western Fresno County, California: U.S. Geol. Survey Prof. Paper 437-A, p. 70.

_____, 1964b, Geomorphology of segmented alluvial fans

in western Fresno County, California: U.S. Geol. Survey Prof. Paper 352-E, p. 89-129.

_____, 1968, Alluvial Fans: J. Geol. Ed., v. 16, p. 101-106.

_____, 1972, Recognition of alluvial fan deposits in the stratigraphic record, in Rigby, J.K., and Hamblin, W.K., eds., Recognition of ancient sedimentary environments: Soc. Econ. Paleont. Mineral. Spec. Pub. 16, p. 63-83.

Burggraf, D.R., 1976, Stratigraphy of the Upper Member, Koobi Fora Formation, southern Karari Escarpment, East Turkana Basin, Kenya (Master of Science Thesis): Iowa State University, Ames, Iowa.

Cerling, T.E., 1977, Paleochemistry of Plio-Pleistocene Lake Turkana and diagenesis of its sediments (Ph.D. Dissertation): Univ. California, Berkeley, California.

_____, 1979, Paleochemistry of the Plio-Pleistocene Lake Turkana, Kenya: Paleogeography, Paleoclimatology, Paleoecology, v. 27, p. 247-285.

_____, Hay, R.L., and O'Neil, J.R., 1977, Isotopic evidence for dramatic climatic changes in East Africa during the Pleistocene: Nature, v. 267, p. 137-138.

Coppens, Y., Howell, F.C., Isaac, G.L., and Leakey, R.E.F., 1976, Earliest Man and Environments in the Lake Rudolf Basin: Univ. Chicago Press, 615 p.

Curtis, G.H., Drake, R.E., Cerling, T.E., Cerling, B.W., and Hampel, J., 1975, Age of the KBS Tuff in Koobi Fora Formation, northern Kenya: Nature, v. 258, p. 395-398.

Eynon, G., and Walker, R.G., 1974, Facies relationships in Pleistocene outwash gravels, southern Ontario; a model for bar growth in braided rivers: Sedimentology, v. 21, p. 43-70.

Findlater, I., 1976, Stratigraphic analysis and paleoenvironmental interpretation of a Plio/Pleistocene sedimentary basin east of Lake Turkana (Ph.D. Dissertation): Birkbeck College, London.

_____, 1978, Stratigraphy, in Leakey, M.G., and Leakey, R.E.F., eds., Koobi Fora Research Project, Vol. 1, Clarendon Press, Oxford, p. 14-31.

Fitch, F.J., and Miller, J.A., 1970, Radioisotopic age determinations of Lake Rudolf artefact site: Nature, Lond.,
v. 226, p. 226-228.

_____, 1976, Conventional Potassium-Argon and Argon-40/
Argon-39 dating of volcanic rocks from East Rudolf, in
Coppens, Y., Howell, F.C., Isaac, G.L., and Leakey,
R.E.F., eds., Earliest Man and Environments in the Lake
Rudolf Basin: Univ. Chicago Press, p. 123-147.

Fitch, F.J., Hooker, P.J., and Miller, J.A., 1976, 40-Ar/39-
Ar dating of the KBS Tuff in Koobi Fora Formation, East
Rudolf, Kenya: Nature, v. 263, p. 740-744.

Frank, H.J., 1976, Stratigraphy of the Upper Member, Koobi
Fora Formation, northern Karari Escarpment, East Turkana
Basin, Kenya (Master of Science Thesis):   Iowa State
University, Ames, Iowa.

Harris, J.W.K., and Isaac, G.L., 1976, The Karari Industry;
early Pleistocene archaeological materials from the
terrain east of Lake Rudolf, Kenya: Nature, Lond., v.
262, p. 102-107.

_____, and Herbich, I., 1978, Aspects of early Pleisto-
cene hominid behavior east of Lake Turkana, Kenya, in
Bishop, W.W., ed., Geological background to fossil man:
Geol. Soc. Lond. Spec. Pub. 6, Scottish Academic Press,
p. 529-548.

Hooke, R.L.B., 1967, Processes on arid-region alluvial fans:
J. Geol., v. 75, p. 438-460.

Hurford, A.J., Gleadow, A.J.W., and Naeser, C.W., 1976,
Fission-track dating of pumice from the KBS Tuff, East
Rudolf, Kenya: Nature, v. 263, p. 738-740.

Isaac, G.L., Leakey, R.E.F., and Behrensmeyer, A.K., 1971.
Archaeological traces of early hominid activities east
of Lake Rudolf, Kenya: Science, v. 173, p. 1129-1134.

_____, 1976, Plio-Pleistocene artifact assemblages from
East Rudolf, Kenya, in Coppens, Y., Howell, F.C., Isaac,
G.L., and Leakey, R.E.F., eds., Earliest Man and Environments in the Lake Rudolf Basin: Univ. Chicago Press,
p. 552-564.

_____, Harris, J.W.K., and Crader, D., 1976, Archaeolo-
gical evidence from the Koobi Fora Formation, in
Coppens, Y., Howell, F.C., Isaac, G.L., and Leakey,

R.E.F., eds., Earliest Man and Environments in the East Rudolf Basin: Univ. Chicago Press, p. 533-551.

Kessler, L.G., 1971, Characteristics of the braided stream depositional environment with examples from the south Canadian River, Texas: Wyo. Geol. Assoc. Earth Sci. Bull., March, p. 25-35.

Leakey, R.E.F., 1970, Early artefacts from the Koobi Fora Area: Nature, v. 226, p. 228-230.

_____, and Harris, J.W.K., 1975, Relationships of Early Pleistocene hominids from Ileret and the Karari Escarpment archaeological sites of the Karari Industry: Am. Assoc. Physical Anthrop. Abs., v. 42, p. 313.

Leakey, M.G., and Leakey, R.E.F., 1978, Koobi Fora Research Project, Vol. 1, The fossil hominids and an introduction to their context: Clarendon Press, Oxford, 191 p.

Leopold, L.B., Wolman, W.G., and Miller, J.P., 1964, Fluvial processes in geomorphology: Freeman, 522 p.

Mathisen, M.E., 1977, A provenance and environmental analysis of the Plio-Pleistocene sediments in the East Turkana Basin, Lake Turkana, Kenya (Master of Science Thesis): Iowa State University, Ames, Iowa.

Miall, A.D., 1977, A review of the braided river depositional environment: Earth Sci. Revs., v. 13, p. 1-62.

_____, 1978, Lithofacies types and vertical profile models in braided river deposits; a summary, in Miall, A.D., ed., Fluvial sedimentology: Canadian Soc. Petr. Geol. Mem. 5, p. 597-604.

Ojany, F.F., and Ogendo, R.B., 1973, Kenya: a study in physical and human geography: Longman Kenya Ltd., 228 p.

Ore, H.T., 1963, Some criteria for recognition of braided stream deposits: Contr. to Geol., v. 3, p. 1-14.

Rust, B.P., 1972, Structure and process in a braided river: Sedimentology, v. 18, p. 221-245.

_____, 1975, Fabric and structure in glaciofluvial gravels, in Jopling, A.V., and McDonald, B.C., eds., Glaciofluvial and glaciolacustrine sedimentation: Soc. Econ. Paleont. Mineral. Spec. Pub. 23, p. 238-248.

Smith, N.D., 1970, The braided stream depositional environment; Comparison of the Platte River with some Silurian clastic rocks, northcentral Appalachians: Geol. Soc. Am. Bull., v. 81, p. 2993-3014.

_____, 1971, Transverse bars and braiding in the lower Platte River, Nebraska: Geol. Soc. Am. Bull., v. 82, p. 3407-3420.

_____, 1974, Sedimentology and bar formation in the Upper Kicking Horse River, a braided outwash stream: J. Geol., v. 81, p. 205-223.

Steel, R.J., 1974, New Red Sandstone floodplain and piedmont sedimentation in the Hebridean province, Scotland: J. Sediment. Petrol., v. 44, p. 336-357.

Vondra, C.F., and Bowen, B.E., 1976, Plio-Pleistocene deposits and environments, East Rudolf, Kenya, in Coppens, Y., Howell, F.C., Isaac, G.L., and Leakey, R.E.F., eds., Earliest Man and Environments in the Lake Rudolf Basin, Univ. Chicago Press, p. 79-93.

_____, 1978, Stratigraphy, sedimentary facies, and paleoenvironments, East Rudolf, Kenya, in Bishop, W. W., ed., Geological background to fossil man: Geol. Soc. Lond. Spec. Pub. 6, Scottish Academic Press, p. 395-414.

_____, and Burggraf, D.R., 1978, Fluvial facies of the Plio-Pleistocene Koobi Fora Formation, Karari Ridge, East Lake Turkana, Kenya, in Miall, A.D., ed., Fluvial sedimentology: Canadian Soc. Petr. Geol. Mem. 5, p. 511-529.

Walker, R.G., 1975, Conglomerate: sedimentary structures and facies models, in Harms, J.C., Southard, J.B., Spearing, D.R., and Walker, R.G., eds., Depositional environments as interpreted from primary sedimentary structures and stratification sequences: Soc. Econ. Paleont. Mineral. Short Course 2, p. 133-161.

Williams, G.E., 1971, Flood deposits of the sand-bed ephemeral streams of central Australia: Sedimentology, v. 17, p. 1-40.

Williams, P.F., and Rust, B.R., 1969, The sedimentology of a braided river: J. Sediment. Petrol., v. 39, p. 649-679.

Williamson, P.G., 1978, Evidence for the major features and development of rift paleolakes in the Neogene of East

Africa from certain aspects of lacustrine mollusc assemblages, in Bishop, W.W., ed., Geological background to fossil man: Geol. Soc. Lond. Spec. Pub. 6, Scottish Academic Press, p. 507-528.

White, H.J., 1976, Stratigraphy of the Lower Member, Koobi Fora Formation, southern Karari Escarpment, East Turkana Basin, Kenya (Master of Science Thesis):   Iowa State University, Ames, Iowa.

Wolman, M.G., and Leopold, L.B., 1957, River flood plains: some observations on their formation: U.S. Geol. Survey Prof. Paper 282-C, p. 87-107.

5. Environments in the Lower Omo Basin from One to Four Million Years Ago

Abstract

For the past four million years, the lower Omo valley has been occupied by a large perennial river draining into a lake. Fossil vertebrates and plants are used to show that climatic conditions were similar to those of the present. The vegetation was characteristic of dry or wooded savanna, and a forest was present along the river. Changes in lake level occurred in response to subsidence or climatic changes at the headwaters of the Omo River.

Introduction

Plio-Pleistocene deposits of the lower Omo valley, in southwestern Ethiopia, have yielded circa 40,000 vertebrate fossils belonging to 125 taxa and including 200 fragmentary specimens of hominids. The stratigraphy of these deposits has been studied in detail (de Heinzelin, et al., 1976), and has been mapped at 1:10,000. This paper describes the principal paleoenvironments as determined by a consideration of the fossil fauna and flora collected from these deposits. Data have been taken from the works of a large number of authors who, in many cases, reached conclusions different than those presented here. The responsibility for errors of interpretation lies solely with the present author.

Geological Background

The depression in which Plio-Pleistocene sediments were deposited in the lower Omo valley was formed at least 4.5 m.y. B.P. In general, the geographic features were much like the present, with the Omo River draining the Ethiopian plateau, and emptying into a lake whose northern shoreline fluctuated from somewhere near its present location to a position at least 100 km north of that. As compared to the present, the Omo River probably had a more linear course, the conflu-

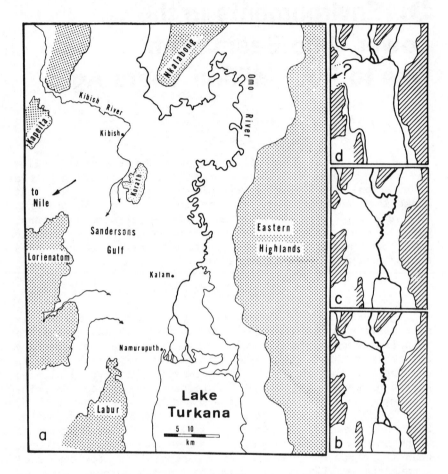

Figure 1. Principal geographic features of (1a) the present
lower Omo valley,   (1b) the schematic paleogeography of the lower Omo valley 3.4 m.y. ago,
(1c) the schematic paleogeography of the lower
Omo valley 2 to 3 m.y. ago, and (1d) the schematic
paleogeography of the lower Omo valley 1.9 m.y.
ago. The northern shoreline of the lake in Figure
1d is near the southern permissible limit.

ence with the Usno River was probably located somewhat further south, and the Korath Range probably did not exist. Figure 1 shows the present geography, and probably topographic configurations at various times in the past.

Three principal formations, the Mursi (140 m; 4.1 m.y.), Usno (170 m; 2.6 to 3.1 m.y.), and Shungura (770 m; 0.8 to 3.7 m.y.) Formations, outcrop in different areas: the Mursi, southeast of the Nkalabong Mts;, the Usno in the northeastern part of the basin; and the Shungura in the southern part of the basin. They are composed of sands, silts, and clays with intercalated tuffs, the latter of fluvial origin. One sequence of sediment types repeated many times begins with coarse sand and grades successively upward into finer sands, silts, and clays, often with carbonate nodules and slicken-sides in the uppermost part. These sequences are 7 to 10 m thick, sometimes truncated at the top, and interpreted as flood plain sediments deposited by a meandering river. In other parts of the section, finely laminated clays and silts predominate and are often accompanied by molluscs and ostra-cods. These are interpreted as being of lacustrine or distal deltaic origin. Deltaic sediments further from the lake are also represented.

A number of subsidiary depositional environments can be discerned within three general types: channel deposits, flood plain deposits, and backswamp deposits (see de Heinzelin, et al., 1976). The presence of Etheria establishes that the river was perennial with depths in channel deposits in the order of 10 to 15 m, indicating that the river had dimensions comparable to those of the present Omo River.

Chronological control has been provided by potassium-argon dates and magnetostratigraphic studies. For details on chronology the reader is referred to Brown, et al., 1978, and Brown and Nash, 1976.

At least eight periods of lake expansion are recorded: two in the Mursi Formation, one in the overlying Nkalabong Formation (Butzer, 1976), one in the Basal Member of the Shungura Formation and its equivalent in the lower part of the Usno Formation, and four in the upper part of the Shungura Formation (in Members G, H, K, and L). The temporal distribution of these is shown in Figure 2, along with a bar graph which illustrates the minimum northward displacement of the shoreline from its present position. A later expansion shown as a single event is recorded by the Kibish Formation of later Pleistocene age, and is also shown on Figure 2. This is not a single expansion, but a record of several high stands detailed by Butzer, et al. (1972). An important

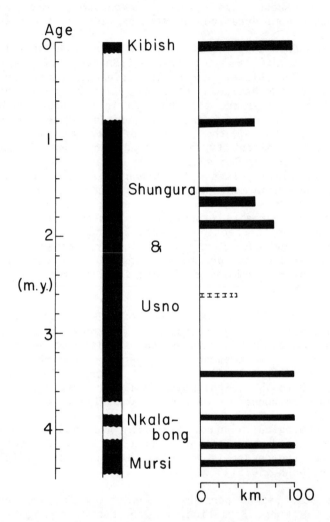

Figure 2. Diagram showing times of expansion of ancient Lake
           Turkana into the lower Omo valley. The bar graph
           indicates the minimum displacement of the shore-
           line northward from that of the present. The cen-
           tral column indicates the formation(s) in which
           these expansions are recorded.

effect of expansion of the lake was the elimination of the vegetation and terrestrial fauna from the lower Omo valley at the indicated times, and the crowding of fauna toward the basin margins.

In summary, normal environments associated with fluvial, deltaic, and lacustrine deposition can be recognized over the entire time period. It is necessary that subsidence took place in the lower Omo valley over the time of deposition of the Plio-Pleistocene formations in order to build up the great thickness (ca. 1 km) of sediments under essentially surficial conditions. The sedimentary record provides some data on the general geography of the region, but for data on the fauna, flora, and climate of the region, we must turn to the fossils.

Fossil Collections and Interpretation

There are five distinct sorts of fossil collections from these formations: invertebrate fossils (largely mollusks collected from lacustrine intervals), microvertebrate fossils, fossil pollen, macrobotanical fossils (wood), and large vertebrate fossils. It is important to note that microvertebrate collections were obtained by sampling at single localities at various stratigraphic levels; they are essentially point localities in space and time. Thus they may record microenvironmental conditions rather than broader environmental conditions. On the other hand, macrobotanical remains and macrovertebrate remains were generally collected from particular stratigraphic levels, but levels which are more broadly defined, and usually from an area of greater lateral extent than was the case for the microvertebrates. In addition, most of the macrovertebrates and all of the macrobotanical remains had undergone some post-mortem transport, and thus represent death assemblages from a number of microenvironments. Therefore they may reflect broader faunal and floral associations. Pollen collections are different again. Like microvertebrate collections, they are essentially point samples in space and time, but the pollen found is interpreted as being composed of a long-distance element, a regional element, and a local element because of the ease of transport of pollen (Bonnefille, 1976). However reasonable the allocation of taxa to these three elements, there is nevertheless an element of ambiguity introduced into the interpretation of the assemblages. For example, Bonnefille (1976) allocated Combretum and Terminalia to the local element (riverine community), but at the generic level these taxa could be as easily referred to the regional element.

There are no marked changes in the molluscan fauna over the time period under discussion which is taken by Gautier (1976) as evidence that the lake was similar in its charac- teristics throughout most of Plio-Pleistocene time, and was fairly fresh at its northern end. However, a thick lacustrine sequence in Member G contains very few mollusks. In this interval, strata were deposited in very shallow water and periodically exposed subaerially, as evidenced by pre- served mudcracks and algal mats. The near absence of mollusks in this part of the section may reflect greater salinity of the lake, the presence of waters which were too turbid, a bottom substrate which was too muddy, or the periodic subaerial exposure of the lake floor.

Microvertebrate collections from upper Member B and Member F of the Shungura Formation include about 35 taxa, and are quite distinct assemblages. The assemblage from Member B has more forest forms (e.g., <u>Pelomys</u>, <u>Eidolon</u>, <u>Suncus</u>), while that from Member F has more forms indicative of a drier environment (e.g., <u>Gerbillurus</u>, <u>Jaculus</u>, <u>Heterocephalus</u>). Jaeger and Wesselman (1976) discuss these differences in detail. Both assemblages, however, contain savanna forms (e.g., <u>Tatera</u>, <u>Paraxerus</u>, <u>Helogale</u>), forest forms (e.g., <u>Taphozous</u>), and riverine forms (e.g., <u>Mastomys</u>, <u>Thryonomys</u>). Thus at both times (ca. 2.7 and 2.0 m.y. ago) there is evi- dence for both savanna and forest forms. By analogy with modern savanna in Africa, we would expect a fringing forest along a permanent watercourse. The microvertebrate collec- tions are consistent with such a diversity of floras. Because only two sites are well-sampled, it is not possible to say whether differences of the regional climate or microenvironmental differences have been documented by the differences in microvertebrate assemblages at the two sites.

Fossil pollen is known for five levels in the Shungura Formation, and complete data have been published for three levels by Bonnefille (1976). In each of these, pollen of Graminae make up 61-74% of the total regional and local pollen, indicating that grass was an important element of the regional flora at each of these times. Bonnefille (1976) states that,

> "The past vegetation of the whole Omo basin was roughly similar to what it is today. There was a forest on the highlands, and a tree shrub sa- vanna on the lowlands. Near the river a wooded vegetation was present, but it appears to have been less developed than the existing riverine forest along the Omo....The mosaic character of the vegetation was already established."

She further states that,

> "About 2 m.y., the vegetation of the Omo basin was
> a savanna with very few trees, which may reflect
> much drier conditions than those which prevailed
> both before and today."

The reader is referred to Bonnefille (1976) for a discussion
of evidence in a pollen spectrum from Member E of the
Shungura formation leading to this latter statement. It is
clear that the pollen spectrum from Member E (ca. 2.1 m.y.)
is very poor in arboreal taxa, but the reasons for the
paucity of arboreal pollen are unclear. Not only is the
regional element poor in arboreal pollen (11 grains), but the
local riverine element is likewise poor (1 grain). The
striking feature of the pollen assemblage from E-4 is that it
it is dominated by three taxa (Graminae, Dasysphaera, and
Polygala) which make up more than 94% of the total pollen.
The two herbaceous genera are forms found today on the grassy
and shrub-covered plains away from the riverine forest of the
Omo, and it is possible that the pollen assemblage reflects a
sample of that local environment rather than widespread
vegetational change.

Fossil wood is fairly common in the Shungura, Usno, and
Mursi Formations, and Deschamps (1976) has identified 58 taxa
from collections made in these formations. Fourteen have
been identified to generic level, and the remainder have been
identified to specific level. These taxa have been divided
into two broad types: those probably not associated with
riverine forest, and those which are possibly associated with
riverine forest. The stratigraphic distribution of these
taxa is shown in Table 1.

The most striking feature of the distribution of macrobotanical taxa is the lack of overlap from one level to
another, and the lack of overlap of fossil taxa with the
modern flora of the lower Omo valley. Some of this lack of
overlap may be because the fossils were collected from different geographic locations corresponding to the outcrop
areas of the formations. The vegetation of the Omo valley
differs in different parts of the valley today, and may also
have differed spatially in the past. For example, Euphorbia
tirucalli is found near the outcrop areas of the Mursi and
Usno Formations, but not in areas near outcrops of the
Shungura Formation. If the macrofossils represent taxa
growing in the area where they were preserved (an assumption
which is easily criticized), then we should not necessarily
expect correspondence of taxa from different geographic
sites. One might also expect a large difference in flora

Table 1. Stratigraphic distribution of fossil plant taxa recovered from the Shungura, Usno, and Mursi Formations.

Taxa possibly associated with riverine forest

	M	LU	A	MU	C	D	F	LG	MG	UG	J	L
Steganotaenia araliacea										x		
Terminalia prunoides									x			
Maytenus senegalensis*									x			
Cathormion altissimum									x			
Zizyphus abyssinica									x			
Acacia nilotica									x			
Brachylaena hutchinsii								x	x			
Lannea weltwitschii									x			
Populus euphratica									x			
Bridelia grandis							x		x			
Ficus capensis							x					
Millettia ferruginea							x					
Trichilia emetica*							x					
Celtis africana							x					
Ficus brachylepsis						x						
Ficus vallis-choudae				x								
Cordia abyssinica				x								x
Erythrophloeum africanum				x								
Borassus aethiopicum				x					x			
Raphia vinifera				x								
Garcinia huillensis			x	x								
Mystroxylon aethiopicum			x	x								
Isoberlinia confusa		x										
Kigelia africana	x											
Ekebergia ruepelliana	x											
Tabernaemontana									x			
Salacia									x			
Macaranga									x			
Landolphia								x	x			
Platysepalum						x						
cf. Myrianthus			x					x	x			
Ficus*			x	x	x	x	x		x			
Garcinia		x	x	x					x			
Diospyros*	x		x					x	x	x	x	x
Baphia	x											

Table 1. continued

Taxa probably not associated with riverine forest

	M	LU	A	MU	C	D	F	LG	MG	UG	J	L
Combretum zeyheri										x		
Ochna schweinfurthiana										x		
Millettia dura										x		
Canthium acutiflorum										x		
Acacia mellifera*										x		
Cola lateritia										x		
Albizia antunesiana							x					
A. adianthifolia							x					
Rothmannia urcelliformis							x					
Garcinia cereoflava						x						
Psorospermum febrifugum					x							
Ozoroa insignis					x		x					
Pseudobersama mossambicensis					x							
Xylia dinklagei				x								
Catophractes alexandri	x											
Stereospermum kunthianum	x											
Combretum collinum	x											
Combretum molle	x									x		
Brachystegia										x		
Commiphora*										x		
Monotes					x							
Indigofera*					x							
Cryptosepalum					x							

---

* These taxa are present in the present flora of the lower
     Omo valley.
**M = Mursi Fm.; LU = Lower Usno Fm.; MU = Middle Usno Fm.;
     A, C, D, F, J, and L = members of the Shungura Fm.;
     LG, MG, UG = Lower, middle, and upper Member G of the
     Shungura Fm.

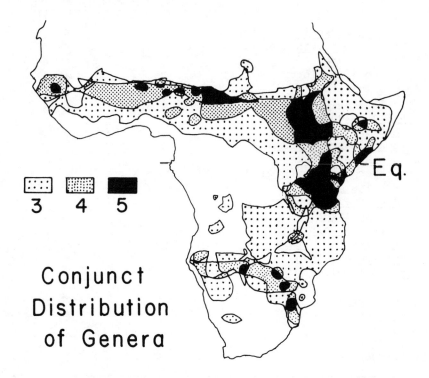

Figure 3. Map of conjunct distribution of genera discussed
           in the text. The numbers refer to the number of
           those genera considered which coexist in each
           of the patterned areas.

following a major lacustrine expansion (e.g., Member G), so floral differences from the top and bottom of this member are understandable. However, there is still a lack of overlap between all members of the earlier, fluvial parts of the Shungura Formation, which must be attributed either to floral changes or to sampling problems. Judging from Carr (1976), the riverine forest is made up of from ten to fifteen dominant woody taxa. There are seven species known from lower Member G, and seven species known from Member C, with one species in common. It is unlikely that there would be this little overlap if we simply drew seven species at random from an identical flora of ten to fifteen species twice, thus it is likely that the flora of the riverine forest changed with time. It is not clear why the riverine flora changed. The species which are documented all have fairly widespread distributions in riverine forest situations and in savanna woodlands today. Those taxa probably not associated with riverine forest are found in a rather broad range of environments ranging from very dry savanna to wooded grassland.

There are about 115 large mammalian taxa which have been recovered from these formations, roughly 100 of which have been identified to generic or specific level (Coppens and Howell, 1976). About two-thirds of the genera are still extant. Nine genera (Elephas, Hippopotamus, Giraffa, Tragelaphus, Kobus, Aepyceros, Hyaena, Papio, and Theropithecus) are found in virtually every member of the Shungura Formation from Member B to Member L, and were most likely present in the area for this entire time period. Most of these are also known in sediments which are older than Member B, but the collections are poor from these earliest levels. One genus, Elephas, is no longer found in Africa. Another, Theropithecus, has a very restricted distribution in the Ethiopian highlands at present, although it is known from Pleistocene deposits over much of sub-Saharan Africa. A third, Aepyceros, is restricted to East Africa at the present time. The remaining genera are widely distributed in Africa, and the geographic distribution of species within these genera has been published as a series of maps by Dorst and Dandelot (1970). From the distribution maps of species, distribution maps of genera were compiled, and from these a map of conjunct distribution of five genera (Giraffa, Tragelaphus, Kobus, Hyaena, and Papio) was constructed (Figure 3). Hippopotamus was excluded because it reflects the presence of a lake or river, and is rather insensitive to other climatic variables. There is essentially no change in the map if Hippopotamus is included. Most other genera recovered from the Shungura Formation are also found within the area defined by these five long-ranging genera. Similar maps were compiled for plant associations, but the regions defined

were much broader. It is significant that the map of conjunct distribution of large mammal genera falls within the
area determined from plant associations, corresponding to
grass savannas and Sudan savannas and woodlands.

Discussion and Conclusions

The geographic regions defined by the generic distribution map have a number of features which are rather well
known. The rainfall is between 40 and 80 cm/yr for the most
part, and roughly half of the year is dry. In the northern
part of the area, the mean annual temperature is 25° to
30° C, and diurnal temperatures range from about 20° to
25° C. These are thought to have been the regional climatic
conditions in the lower Omo valley for the period from one to
four m.y. ago. Those plant and animal taxa which require
more mesic conditions are thought to have lived along the
river where a ready supply of water was available. In
summary, it is felt that there is no conclusive evidence for
major climatic change in the lower Omo basin over this time
period. Such changes as may have occurred were most likely
small in magnitude so that the vegetation may have been
similar to that characteristic of dry savannas at times, and
similar to that characteristic of wooded savannas at others.
In all likelihood, neither extensive forest nor desert
conditions ever prevailed.

Apparent changes in the level of ancient Lake Turkana
may have been caused either by tectonic movements, or by
changes in climate. The former might involve subsidence of a
portion of the basin, a change in the area of the lake, or
the diversion of a river draining into the lake (Cerling,
1979). Changes in climate involve greater or less evaporation from the lake surface, or changes in the inflow of water
to the lake. Most of lacustrine intervals are thin: a few
meters to a few tens of meters in thickness, and these are
thought to have been caused by climatic changes. Earlier
(Brown, et al., 1978) it was argued that all lacustrine
intervals were of climatic origin, but the thick lacustrine
sequence in Member G at about 1.9 m.y. B.P. may have been
caused by tectonic movements. This interval consists of
about 90 m of lacustrine deposits which have a very sharp
contact with the underlying fluvial deposits. There is no
interval of transition; rather it appears that the northern
shoreline of the lake was displaced northward very rapidly.
The strata in this interval were deposited in shallow water,
and were periodically exposed. The sequence ends with the
return of deltaic and fluvial sedimentation. There is no
evidence of incision at the top of the lacustrine sequence
which would be expected if lake level were lowered in

relation to the sedimentary surface. Therefore it is thought that this interval represents an increase in the rate of subsidence of the basin so that deltaic sedimentation was displaced northward. Regression was most likely the result of decrease in the rate of subsidence so that the river was able to build its delta southward again.

While it has been argued that climatic changes were not of very large magnitude in the lower Omo basin, most of the water in Lake Turkana is derived from the Ethiopian highlands and the highlands of Kenya and Uganda. Changes in rainfall there would affect lake levels even if the local climate were only slightly changed. The lake may have acted as a recorder of climatic events which took place far from its own geographic location.

It should be noted that the Omo River allows water-dependent species to live in the lower Omo valley in habitats along its flood plain. Because the Omo River is also the principal source of sediments, those species which live on the flood plain are the ones most likely to be preserved as fossils. However, their fossil record will reflect more mesic conditions than those of the surrounding area. This is a general problem with sediments deposited by large rivers which flow into regions more arid than the regions from which they arise. Chapman and White (1970) have pointed out that,

> "Nearly all (at least 95%) of the Guineo-Congolian species occurring in Malawi in evergreen forests on well-drained sites also occur in fringing forests in the Sudano-Zambezian region, and most of them are widely distributed there."

Therefore, fossil plants representative of forests may only reflect the presence of a large river, and not forested conditions. Further, riverine plants are those most likely to be preserved, which might lead to paleoclimatic interpretations which err on the wet side. Consequently, more weight should be given to plant and animal taxa which are independent of water when attempting regional climatic analysis from fossil assemblages.

In conclusion, the lower Omo valley has been occupied by a large perennial river bordered by forest which flowed southward into a lake for the past 4.5 m.y. The area surrounding the river was rather dry, receiving between 40 and 80 cm of rainfall per year. The floral composition of the riverine forest and surrounding region changed in time, but there is no clear trend toward either much wetter or much drier conditions than those which presently prevail in the area. There are many phytogeographical problems which are raised by the fossil plant taxa which have been recovered, but all taxa can be accommodated in the Sudano-Zambezian region.

References

Bonnefille, R., 1976, Palynological evidence for an important change in the vegetation of the Omo Basin between 2.5 and 2 million years, in Coppens, Y., Howell, F.C., Isaac, G.L., and Leakey, R.E.F., eds., Earliest Man and Environments in the Lake Rudolf Basin: Univ. of Chicago Press, Chicago., pp. 421-431.

Brown, F.H., and Nash, W.P., 1976, Radiometric dating and tuff mineralogy of Omo Group deposits, in Coppens, Y., Howell, F.C., Isaac, G.L., and Leakey, R.E.F., eds., Earliest Man and Environments in the Lake Rudolf Basin: Univ. of Chicago Press, Chicago., pp. 50-63.

Brown, F.H., Howell, F.C., and Eck, G.G., 1978, Observations on problems of correlation of late Cenozoic hominid-bearing formations in the North Lake Turkana Basin, in Bishop, W.W., ed., Geological Background to Fossil Man: Scottish Academic Press, Edinburgh, pp. 473-498.

Brown, F.H., Shuey, R.T., and Croes, M.K., 1978, Magnetostratigraphy of the Shungura and Usno Formations, southwestern Ethiopia: new data and comprehensive reanalysis: Geophys. Jour. Roy. Astr. Soc., 54, 519-538.

Butzer, K.W., 1976, The Mursi, Nkalabong and Kibish Formations, lower Omo Basin, Ethiopia, in Coppens, Y., Howell, F.C., Isaac, G.L., and Leakey, R.E.F., eds., Earliest Man and Environments in the Lake Rudolf Basin: Univ. of Chicago Press, Chicago., pp. 12-23.

Butzer, K.W., Isaac, G.L., Richardson, J.L., and Washbourn-Kamau, C., 1972, Radiocarbon dating of East African lake levels: Science, 175, 1069-1076.

Carr, C.J., 1976, Plant Ecological variation and pattern in the lower Omo basin, in Coppens, Y., Howell, F.C., Isaac, G.L., and Leakey, R.E.F., eds., Earliest Man and Environments in the Lake Rudolf Basin: Univ. of Chicago Press, Chicago., pp. 432-470.

Cerling, T.E., 1979, Paleochemistry of Pli-Pleistocene Lake Turkana: Paleo Paleo Paleo., 27, 247-285.

Chapman, J.D., and White, F., 1970, The evergreen forests of Malawi, Oxford: Commonwealth Forestry Institute, Oxford, 190 p.

Coppens, Y., and Howell, F.C., 1976, Mammalian faunas of the Omo Group Distributional and Biostratigraphic aspects, in Coppens, Y., Howell, F.C., Isaac, G.L., and Leakey, R.E.F., eds., Earliest Man and Environments in the Lake Rudolf Basin: Univ. of Chicago Press, Chicago., pp. 177-192.

de Heinzelin, J., Haesaerts, P., and Howell, F.C., 1976, Plio-Pleistocene Formations of the lower Omo Basin, with particular reference to the Shungura Formation, in Coppens, Y., Howell, F.C., Isaac, G.L., and Leakey, R.E.F., eds., Earliest Man and Environments in the Lake Rudolf Basin: Univ. of Chicago Press, Chicago., pp. 24-29.

Deschamps, R., 1976, Resultats preliminaires de l'etude de boit fossiles de la basse vallee de l'Omo (Ethiopie sud-occidentale): Mus. Roy. Afr. Centr., Tervuren (Belge). Dept. Geol. Min. Rapp. Ann., 1975., pp. 59-65.

Dorst, J., and Dandelot, P., 1970, A field guide to the larger mammals of Africa, Collins, London, 287 p.

Gautier, A., 1976, Assemblages of fossil freshwater mollusks from the Omo Group and related deposits in the Lake Rudolf Basin, in Coppens, Y., Howell, F.C., Isaac, G.L., and Leakey, R.E.F., eds., Earliest Man and Environments in the Lake Rudolf Basin: Univ. of Chicago Press, Chicago., pp. 383-401.

Jaeger, J.J., and Wesselman, H.B., 1976, Fossil remains of micromammals from the Omo Group deposits, in Coppens, Y., Howell, F.C., Isaac, G.L., and Leakey, R.E.F., eds., Earliest Man and Environments in the Lake Rudolf Basin: Univ. of Chicago Press, Chicago., pp. 351-359.

James L. Aronson, Maurice Taieb

6. Geology and Paleogeography of the Hadar Hominid Site, Ethiopia

Abstract

The Hadar hominid site occurs in the western Afar depression of Ethiopia and is underlain by the Hadar Formation. The site occurs in the large basin that accumulated sediments from late Miocene to early Pleistocene time between the Ethiopian Escarpment and the main volcanic axis in central Afar. The Hadar Formation is middle Pliocene in age, dating from 2.5 to 4.0 m.y. B.P. The formation has a variable thickness of about 200 m. It contains an abundant well-preserved vertebrate fauna in a cyclical sedimentary sequence dominated by thick non-laminated silty clay units with thinner laterally extensive sands. The sands are fluvially deposited, probably by braided rivers in a distal or delta-plain setting where the thicker silty clay units could have accumulated as overbank deposits on the low interfluvial or inter-distributary plain. A lake existed east of the plain and transgressed across the Hadar site three times during the deposition of the Hadar Formation. Faulting, hypothesized to have been active during deposition, may have caused uplifted surfaces to undergo soil formation during deposition of the Formation.

The sedimentary sequence at Locality 333, the most prolific fossil early hominid site known anywhere, illustrates the shifting of the above environments across the Hadar site.

Introduction

The Hadar hominid site is named for the prominent Pliocene fossiliferrous strata exposed in a badlands topography along the Kada Hadar and Ounda Hadar tributary wadis to the Awash River in the Afar depression. The site was discovered in 1968 (Taieb, 1974). After preliminary

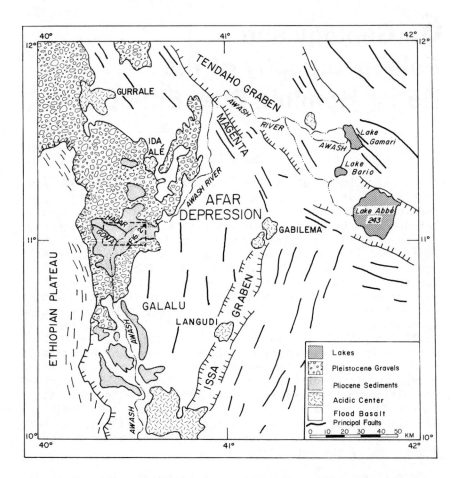

Figure 1. Regional Map of the Western Afar Depression,
showing the Hadar Site along the Awash River and
the Miocene-mid-Pleistocene sediments of the west
central Afar sedimentary basin. Flood basalts of
the central Afar and rhyolitic centers are shown.
The principal line of tensional tectonic and
volcanic activity is the Issa-Tendaho graben. The
Pleistocene gravels contain Acheulean artifacts.

reconnaissance of the region indicated that the area at Hadar was particularly rich in vertebrate fossils (Taieb et al., 1972), the International Afar Research Expedition (IARE) was mounted. Remarkable hominid fossil finds by D.C. Johanson, co-leader of the expedition, began during the first field season in 1973 and were capped in 1974 with the recovery of an associated partial skeleton, nicknamed Lucy, the most complete early hominid known. In the succeeding three-month field seasons of 1975 and 1976, Afar Locality 333 (A.L. 333) was discovered and excavated with its remarkable preservation of at least 13 hominid individuals together. Pliocene artifacts were discovered in 1976 at Gona to the east of Hadar (Roche and Tiercelin, 1977), but their age relations are still under study. The Ethiopian field studies temporarily ceased after the 1976 season, but should commence again in late 1981.

Geologic studies at Hadar have not kept pace with the unusually prolific fossil recovery as they require considerable field time and laboratory study. Two current studies will aid paleoenvironmental reconstruction by examining paleoecological communities (Gray, in progress) and by sedimentology (Tiercelin, in progress). Preliminary interpretations of area sedimentology and environment of deposition have been published by Taieb and Tiercelin (1979, in press).

Location

The area which has been explored thus far is only 16 km by 6 km. It is located at 11°N latitude and 40.5°E longitude in the Miocene-Quaternary west central Afar sedimentary basin (Taieb et al., 1976). This sedimentary basin is an ill-defined strip of up to 60 km wide extending about 150 km from 10°N to 11-1/2°N latitude (Fig. 1). The basin lies on the very western edge of the Afar depression between the Ethiopian Escarpment on the west, and the main Afar region of block-faulted flood basalt on the east. Beginning in the late Miocene, this elongate fault-bounded trough has been a sediment trap involving an interplay of lacustrine and fluvial processes.

Most of the Miocene-Pliocene sediment fill in this vast sedimentary basin is unavailable for exploration, as it is covered by a veneer of up to 10 m of Pleistocene gravels. These latter are partly alluvial and partly of residual lag origin. They represent the distal part of coalescing piedmont fans that extended from the Ethiopian Escarpment in the middle and late Pleistocene. It is probably only recently that the modern Awash River has occupied or reoccupied the region. The lowering of its base-level in the central Afar

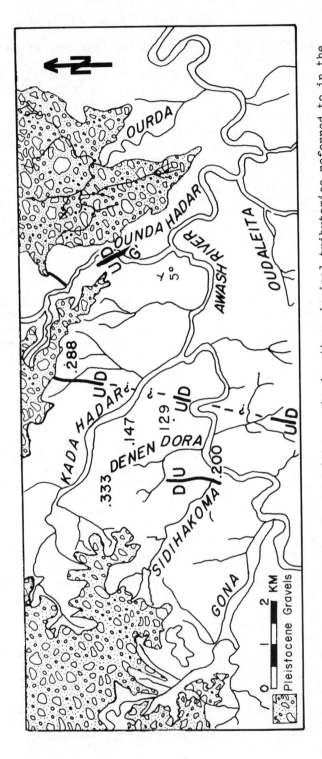

Figure 2. Map of the Hadar Hominid Site, showing the principal tributaries referred to in the text, the Pliocene sedimentary strata and the Pleistocene capping gravels. The three principal faults in the Pliocene section are shown. The fault marked G is a relatively small one which is known to have been active during deposition. Localities 129, 147, 200, 288, and 333 are discussed in the text.

east of the Magenta fault scarp (Fig. 1) has caused the river to cut down through the Pleistocene capping gravels and into the older sediments. At Hadar, tributaries like the Gona, the Kada Hadar, and the Ounda Hadar (Fig. 2) are rapidly eroding headward to form a badlands topography of about 150 m in the unconsolidated Pliocene sediments. The rapid erosion is accentuated by the particular desert climate with seasonal flash storms and little vegetative cover.

Much of the fossil material at Hadar has been recovered from float as lag material on the slopes. Most of the fossils can be placed fairly accurately into their original stratigraphic position because the intricate gulley and ridge exposure usually confines each slope to a narrow stratigraphic interval upslope of a find.

Tectonic Setting

The Afar is the triple junction of the East African Rift, the Red Sea Rift and Gulf of Aden Rift. The extreme tensional tectonics have resulted in crustal extension similar to that of Iceland. This has occurred by thinning the crust and/or by actual basaltic crust generation (Barberi et al., 1972). Regional age determinations of the flood basalts at the latitude of Hadar by R.C. Walter (in preparation) show a progressive increase in age going west from central Afar to the Ethiopian Escarpment. A similar pattern was previously indicated at 13°N latitude (Barberi et al., 1972). Proceeding west from central Afar to the Hadar part of the sedimentary basin the ages of the flood basalt progressively increase in a general way from 1.5 to 2 m.y. at the Magenta fault scarp (Fig. 1) to 3.5 to 4.0 m.y. at the eastern edge of the basin. On the western side of the basin at the latitude of Hadar, basalts in the step-like frontal fault blocks of the Ethiopian Escarpment range from about 3 to 5 m.y. This pattern is consistent with a spreading process which has acted since about 4 or 5 m.y. B.P. to create a volcanically active area east of the Ethiopian Escarpment (Aronson et al., in press). The axis of volcanism has progressed eastward through time to its present location leaving older basalt in its wake. The volcanically active line today is along the Issa and Tendaho grabens (Fig. 1) whose bounding faults on the west raise the Afar flood basalt up to form the west-sloping Galalu Plain and Magenta Plateau. Both of these west-tilting highlands in central Afar have acted to confine drainage and thereby trap sediment derived from the Ethiopian Escarpment in the west central sedimentary basin.

The 4 m.y. old basalts on the Ethiopian Escarpment and the 2 m.y. old basalts on the Magenta Escarpment are cut and

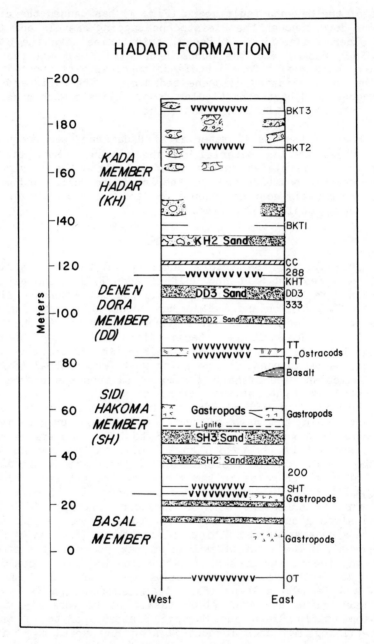

Figure 3. Generalized geologic column for the Hadar Formation showing the principal sand units, tuffs, and
marker horizons. The four members are bounded
isochronously by tuffs.

uplifted by major faults. Thus the relief today, particularly at the Ethiopian Escarpment, conceivably may not have been so pronounced in late Miocene through middle Pliocene time.

Several young large central rhyolitic volcanoes are spaced along the presently active spreading line in the Issa and Tendaho grabens. According to Walter (in preparation) a similar old eroded central rhyolitic complex only 40 km northeast of Hadar at Ida Ale was active from 3.2 to 3.6 m.y. B.P., surrounded by comparably aged flood basalt. This is consistent with the possibility of the central Afar spreading axis having been much closer to Hadar during the Pliocene.

A major unknown in determining the tectonic setting of the west central sedimentary basin is the thickness of the sedimentary deposits, and the nature and age of the basin floor. Because flood basalts in the range of 3.5 to 4 m.y. old occur in either flank of the basin at the latitude of Hadar, it seems reasonable to project the occurrence of down-faulted equivalents of these or older basalts to be the basin floor at Hadar.

Stratigraphy of the Hadar Formation

Taieb has provided a coherent stratigraphic organization of the Hadar Formation that is workable and easy to follow in the field (Johanson et al., 1978). The idealized geologic column is shown in Figure 3. The formation is defined only on a working basis, because its base and top have not been defined. The formation is subdivided into four members from the bottom to the top: the Basal Member, the Sidi Hakoma Member, the Denen Dora Member, and the Kada Hadar Member. Proceeding from bottom to top, these members are separated from each other by three isochronous tuffs: the Sidi Hakoma Tuff (SHT, Fig. 3), which is a ubiquitously exposed double layer tuff; the Triple Tuff (TT, Fig. 3), a series of three or four 2 to 5 cm thick tuffs which is also ubiquitously exposed in an extensive 10 m thick sequence of mudstones; and the Kada Hadar Tuff (KHT, Fig. 3), which is only discontinuously preserved, having been eroded in many places prior to deposition of the overlying sediment. The majority of the fossil vertebrates and all of the hominid fossils are derived from the Sidi Hakoma, Denen Dora and lowermost Kada Hadar Members.

A key point is that the thickness of the formation seems to vary considerably over the Hadar area from about 150 to 250 m. This variation occurs principally by increases in thickness in the Sidi Hakoma Member (Fig. 4) going from west to east.

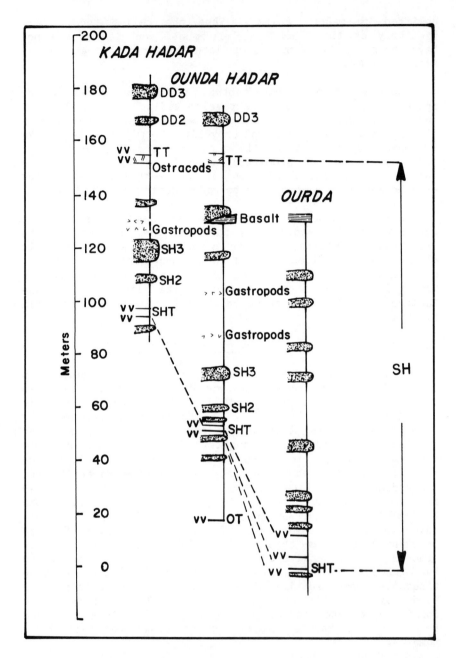

Figure 4.   Stratigraphic correlations by Taieb and Tiercelin,
in press, showing the major increase in thickness
that occurs in the Sidi Hakoma Member going east
across the Hadar Site.

In a general sense the Hadar Formation consists of interbedded, alternating beige sand units and dark brown silty clay units, both of which are fairly continuous laterally. The sand units are about 2 to 8 m thick and the silty clay units about 10 to 40 m thick. Overall, the sand units make up only about 20 percent or less of the formation. The relatively regular interbedding and the degree of lateral continuity of many of the units gives the impression of rhythmic sedimentation.

Taieb has further subdivided the individual members of the formation by means of both recognizable marker beds (other than the tuffs), and by means of laterally continuous or partially continuous sands (Johanson et al., 1978) (Fig. 3). For example, the Sidi Hakoma Submember #2 Sand delineates the base of that submember. The sand is laterally continuous across most of the area. The top of this submember is the base of the Sidi Hakoma Submember #3 Sand. These sands are almost certainly time-transgressive units, unlike the isochronous tuffs. Some of the non-tuff marker horizons are uniform and widespread and are possibly of lacustrine origin. They are likely to be approximately isochronous. These include a lignite layer, several thin beds of densely packed gastropod shells, and two fissile sub-laminated clay horizons which are excellent site-wide markers. Of these latter two clay marker horizons, the lower one is the Triple Tuff Ostracod Clay, with an ostracod-rich zone in the middle of it. It occurs in the lower part of the Denen Dora Member. The other is the Confetti Clay in the lower Kada Hadar Member. In addition to these markers are the three Bouroukie Tuffs (BKT 1, 2, 3 in Fig. 3) used to subdivide the Kada Hadar Member and the two Oudaleita Tuffs (O.T. in Fig. 3) which occur near the stratigraphically lowest exposures of the Basal Member.

Provenance of the Hadar Sediments

The provenance of the Pliocene sediments at Hadar can be determined through a study of the mineralogy and petrography of the sands and conglomerates. Results from just a few samples of the sands are presently compiled. Their analysis gives further insight into the tectonics of the Hadar basin. These sands and conglomerates are distinctly volcaniclastic. Point-counting of typical Hadar sand units (Sidi Hakoma Submember #2 Sand, Denen Dora Submember #3 Sand and Kada Hadar Submember #2 Sand) shows that 80-90% of the grains in each are volcanic rock fragments, with acidic types being a little more common compared to basaltic types. Typically plagioclase, ferromagnesian minerals, and quartz each represent only 2-5% of the grains. Doubly terminated pyramidal grains of the high temperature volcanic form of quartz are present. All together this assemblage indicates the Pliocene Hadar sands are derived by means of relatively little sedimentary reworking of volcanic detritus from a volcanically active area that was close to the basin. A tectonic setting comparable to the present is implied with the possible exception that rhyolitic central volcanoes were a larger component of the source region then, than now. These results on the sands are confirmed by brief field examination of cobble clasts from channel conglomerates from Sidi Hakoma Submember #3 Sand, Denen Dora Submember #3 Sand and Kada Hadar Submember #2 Sand. These clasts are entirely composed of volcanic rocks. Most of these are judged to be andesite to rhyolite in composition. There is a minor component of weathered basalt.

A few K-Ar ages have been measured on trace constituents of sanidine from selected sands giving ages which range from about 4.5 to 9.0 m.y. B.P. (Aronson et al., in press). This further confirms that a young volcanically active tectonic source region surrounded Hadar during the Pliocene, just as it does today.

Age of the Hadar Formation

Most of the original tuffs at Hadar have been altered to clay and are unsuitable for geochronology. Most recent radiometric age data for the formation indicate the Bouroukie Tuff 2, which is a crystal-bearing primary air-fall tuff (Aronson et al., 1977) is about 2.85 m.y. old. This tuff, located in the middle of the Kada Hadar Member, overlies all the fossil material and thus sets a minimum age to the fossils. We are investigating the possibility of its correlation with one of the tuffs just below the artifact site immediately east of Hadar at Gona (Fig. 2). In addition to the Bouroukie Tuff 2, some geochronologic control comes from a basalt in the Sidi Hakoma Member. The Kada Moumou basalt flow from the east forms an intertonguing relationship with the upper Sidi Hakoma Member. Its age is uncertain because alteration of its glassy matrix has apparently altered the K-Ar ratios of most of the samples dated so far. As a result, we have obtained a variation in measured ages that correlates with the degree of alteration. The least altered and finest grained sample of the flow we have been able to find since our first report (Aronson et al., 1977) yields a probable minimum age measurement for the flow of 3.6 to 3.7 m.y. B.P. This compares to our previously reported oldest ages of about 3.1 m.y. B.P. (Aronson et al., 1977). In addition to the less altered condition of the new sample, part of the increase in the age measurement is due to use of the new decay and abundance constants for ^{40}K, and to the use of defined values for the Case Western ^{38}Ar internal standard compared to other labs.

The fossil terrestrial vertebrate fauna in the Hadar Formation is especially abundant and of good quality for establishing a biostratigraphic age for the site. At present the best correlation with other East African sites where fossil evolutionary trends have been established occurs via the fossil pigs (Cooke, 1978; Harris and White, 1977). A key fossil in this regard is <u>Nyanzachoerus</u> <u>pattersoni</u> (Cooke, 1978) (also referred to as <u>Nyanzachoerus</u> <u>kanamensis</u> (Harris and White, 1977). It is present only in the lower two members whereas another biostratigraphically important pig, <u>Notochoerus</u> <u>euilus</u>, occurs throughout the formation (Gray, pers. comm.). In the rest of East Africa the <u>Nyanzachoerus</u> genus became extinct about 3 to 4 m.y. B.P., with one species of the genus evolving into <u>Notochoerus</u> <u>euilus</u>, during that time (Harris and White, 1977). The <u>Notochoerus</u> <u>euilus</u> in the Sidi Hakoma Member at Hadar is considered to be primitive (Harris and White, 1977) and compares well with <u>Notochoerus</u> at Laetoli (Harris and White, 1977) which has been radiometrically dated at 3.7 to 3.8 m.y. B.P. (Leakey <u>et al.</u>, 1976). The occurrence of <u>Elephas</u> <u>ekorensis</u> and <u>Loxodont</u> <u>adaurora</u> in the Sidi Hakoma and Denen Dora Members suggests a correlation with Member A and the lower part of B at Shungura and thus a pre-3.0 m.y. B.P. age (Beden, 1979).

The magnetic polarity reversal record at Hadar seems to be fairly regularly developed. It is being scrutinized in order to possibly refine the age assessment and to test ideas on the time-transgressive character of sedimentary layers deposited during polarity transitions.

Structural Geology and
Possible Controls on Sedimentation

The major structural feature at the Hadar area is a rather abrupt change in attitude of the beds within the confines of the site. At Kada Hadar and to the west, the beds are flat-lying, but in Ounda Hadar the sediments assume a pronounced 5° northeast dip. Several small faults cut the formation, but three faults have been detected with significant stratigraphic throw. The three faults, all with north northeast strike, are shown in Figure 2. They dip steeply with a slip that is dominantly normal and thus due to tensional forces. The eastern and western faults have stratigraphic throws of only about 10 m, down to the west. The central fault, known as the Locality 288 Fault, has a throw of 40 m down to the east at Locality 288. The fault probably connects with observed segments on-strike to the south (Fig. 2). However, the throw on these segments is only a few meters. If this connection is valid it would mean that the overall motion of the Locality 288 Fault has been clockwise rotational. Such rotation east of the Locality 288 Fault explains the regional change in dip that occurs between the Kada and Ounda Hadar.

A major geologic problem at the Hadar area is the increase in thickness of the Sidi Hakoma Member as one moves east about 4 km from Kada Hadar to Ounda Hadar. This isochronously bounded member is 60 m thick at Kada Hadar, but is 100 m thick at Ounda Hadar. This trend intensifies further to the east where the member becomes 150 m thick only 2 km further east again at Ourda (Taieb and Tiercelin, in press) (Fig. 4). The change in the thickening is tentatively attributed to activity on the part of a Locality 288 fault-progenitor and other faults during the deposition of the Sidi Hakoma Member. Such structures could have been intermittently expressed on the Pliocene depositional surface as fault scarps that interrupted sedimentation on the relatively upthrown blocks to the west. The even greater thickening of the Sidi Hakoma Member further east at Ourda demands the presence of a second fault, or faults, between Ounda Hadar and Ourda that has not been explored sufficiently.

So far there are only two pieces of evidence supporting such "growth fault" activity at Hadar. The Locality 288 Fault underwent much of its 40 m displacement prior to deposition of the overlying Pleistocene gravels, but did continue activity after their deposition so as to displace the gravels 5 to 10 m at locations north of Locality 288. Also a small fault in the Ounda Hadar (labeled G in Fig. 2) displaces the Sidi Hakoma Submember #3 Sand about 5 m but does not displace strata about 10 m stratigraphically above this sand. The displacement is down to the west.

Sediment of the Hadar Formation

A. <u>Sands</u>. The sand units commonly consist of medium to coarse sub-rounded grains. The thicker units are 3 to 8 m thick and are partially cemented with calcite. Where cemented they are erosionally resistant enough to hold up benches in the topography. A few of the sand units (Sidi Hakoma Submember #2 and #3 Sands; Denen Dora Submember #3 Sand; and Kada Hadar Submember #2 Sand) can be traced continuously across all or most of the exposed Hadar area and appear like blanket sands on that scale. Typically these beds have moderately flat basal contacts with occasional

1-m-deep by 5-m-wide scours. Additionally, major channel fills, a few meters thick and 15 m wide, have been observed at the base of some of the sands in a few locations.

The sands exhibit structures indicating a fluvial origin. Trough cross-bedding predominates with sets 1 to 3 m in breadth and 50 cm in thickness.

A regular fining-upward sequence has been demonstrated for some sands in only a few localities. For example, in the Denen Dora #3 Sand in the Gona area (Fig. 2) a pebble basal scour is succeeded by trough cross-bedding and then fine sand and silt with abundant root casts. Trough-shaped scours are occasionally filled with granule conglomerates of calcrete and mudstone clasts. However, at other places a root-cast zone occurs in the middle of a sand bed and a simple fining-upward sequence does not exist.

In addition to the blanket sand horizons, other laterally extensive sand bodies occur. Where observed, the edge of these sand bodies is in a steeply curved lateral contact with silty mudstone that encases the sand body. The sand bodies called the Denen Dora Submember #2 Sand are of this type, as is the Kada Hadar sand below the Confetti Clay at Locality 288 which is believed to have originally contained the hominid skeleton.

The final category of sand body is a simple isolated channel fill. These channels are either tabular and about 50 cm deep by 10 to 15 m wide, or "U"-shaped and several meters deep. Examples of the tabular channel fill occur below the Denen Dora Submember #3 Sand of Locality 333. Two of the "U"-shaped channels that are particularly evident were developed and filled at the time of the eruptions of Sidi Hakoma Tuff and Kada Hadar Tuff respectively. These fills are entirely or largely filled with volcanic glass shard. In the case of the Sidi Hakoma channel the paleo-land surface that existed to either side of it and which was mantled by the 0.5-to 1-m-thick tuff was a very flat surface. The 150 m of channel length exposed is remarkably straight. The fill is medium-scale trough cross-bedded and the current direction for the channel bears north. Further in an up-current direction, two channels are exposed. Still further up-current on the south side of the Awash a broad basin existed that was filled with laminated tuff followed by trough cross-bedded tuff 6 m thick (Taieb and Tiercelin, in press).

The existence of isolated channel fills in the Hadar sequence and the straightness of some of the channels may indicate that accumulation of Pliocene sediment at Hadar was occasionally interrupted by non-deposition accompanied by channelized erosion.

Significant amounts of gravel with cobbles 7 to 12 cm in diameter first occur locally as a broad paleo-channel fill in the Sidi Hakoma Submember #3 Sand at Sidi Hakoma. All of the gravel occurrences are clast-supported. Some minor gravel lenses occur in the other sands at the base of scours. Significant changes occurred at the Hadar site by middle Kada Hadar time as gravels become common units in this part of the section. In general the sand units of the Kada Hadar Member become more conglomeratic both upward in a stratigraphic sense and westward toward the Ethiopian Escarpment in a directional sense. This involves both the number and thickness of gravel units, the extreme being an individual conglomerate 40 m thick that is stratigraphically high in the Kada Hadar Member in the Gona area (Taieb and Tiercelin, in press).

One sand horizon in the lower Kada Hadar Member just above the Confetti Clay marker bed was followed in detail through the western part of the Hadar site in order to observe the transition of it to a gravel bed. This sand unit is about 3 m thick and is encased in a mostly dark brown silty clay. The sand first appears between the Gona and Sidi Hakoma tributaries as a fine-grained sand with low-angle planar cross-beds. It is in lateral contact with the silty clay. Progressing over a few kilometers to the west, this fine sand coarsens and begins to contain scattered pebbles. These become coarser and more abundant until at Gona the unit is entirely a thick sheet of cobble gravel still encased in brown silty clay. This sheet gravel persists for another 2 km further west, the limit of exploration.

Terrestrial vertebrate fossils are very abundant locally in the sand units up through the lower Kada Hadar Member. Indeed the sands have supplied probably 80% of the Hadar fossil vertebrates. Most of this material is disarticulated and consists of heavier cranial pieces, or post-cranial material from larger animals. Notable among the fossil vertebrates are the elephants, pigs, hippos, bovids, horses, and rhinos. Also common in the sands are 1- to 1.5-m-wide carapace of giant turtles and skull material of crocodiles. There are several clutches of a dozen or more crocodile eggs and of turtle eggs that occur in the sand bodies, particularly in Sidi Hakoma Submember #3 Sand and Denen Dora Submember #3 Sand. In addition to vertebrate fossils, the sands occasionally contain accumulations of silicified twigs and branches of wood.

B.   Silty Clays.   These argillaceous units constitute
the majority of the Hadar Formation. They are beige if they
contain significant silt, and dark brown, maroon or green if
they do not. Except for the Confetti Clay marker clay and
parts of the Triple Tuff marker clay, the Hadar silty clays
are typically massive with either slickensided curved fractures or blocky fractures.   The two marker clays, on the
other hand, are slightly fissile and faintly laminated.
Calcite concretions are common in the massive silty clays.
These are often confined to fairly narrow, decimeter-thick
zones but in other cases they are sparsely scattered.   In
some cases the concretions can be smooth and round, about 3
cm in diameter, whereas in other units the concretions are
large and irregular-shaped and even merge to form continuous
limestone or limestone-rich zones 5 to 10 cm thick.

The Confetti Clay marker bed is unfossiliferous whereas
the Triple Tuff marker clay is distinguished by a laterally
persistent zone, 1 to 3 cm thick that is rich in ostracod
tests, scales, bone fragments and teeth of fish, and occasional whole gastropod shells.

Many of the silty clay units that are not laminated have
laterally persistent zones with abundant gastropods and occasional pelecypods dispersed over a thickness of 1 to 2 m.
The sparser vertebrate fossil fauna in the silty clay units
compared to the sand units is also biased either toward cranial material or toward post-cranial material from the larger
vertebrates.   The silty clay sequence that encompasses Denen
Dora Submember #1 is noteable for its abundance at a few
localities of horn cores of the bovid, Kobus (waterbuck)
(Gray, pers. comm.).   There are a few occurrences of associated vertebrate skeletons recovered from the silty clays.
These include one complete articulated elephant skeleton from
Sidi Hakoma Submember #4, a complete articulated skeleton of
an alcelaphine bovid from the upper part of the Denen Dora
Submember #3, and a complete articulated horse skeleton from
Denen Dora Submember #2 (Gray, pers. comm.).   Many of the
important hominid fossils are also recovered from silty
clays.

C.   Other Units.   In addition to the silty clays and
sands which comprise the bulk of the formation, three other
less abundant sediment types are important as they indicate a
shallow lacustrine or marshy peri-lacustrine setting had
developed at Hadar during their deposition. These sediment
types are the lignite, the gastropod shell layers, and a
stromatolitic limestone. The lignite occurs as a laterally
extensive 5- to 10-cm-thick layer rich in plant fragments in
the middle part of the Sidi Hakoma Member. There are about

five gastropod shell layers which occur in the upper Basal Member and in the Sidi Hakoma Member. Typically these are 10 to 20 cm thick, and composed of gastropod shells in a matrix of coarse silt. Two horizons in the Upper Sidi Hakoma Member are so dense in shells that they act as resistant limestone units. Almost all the shells are whole and dominantly comprised of two genera, Melanoides and Bellamya. The shells represent a life assemblage with individuals ranging from 2 to 30 mm in length, deposited with little or no preferential alignment. In some cases the shells are propped up against each other. Along with the gastropods are common large, heavy-walled, smooth, unionid, pelecypod shells.

The stromatolite limestone occurs at Locality 147 over an area 200 m wide. It consists of inter-connected patches and lenses of stromatolitic limestone in the Upper Sidi Hakoma Member between the lignite and the two dense gastropod zones. The unit is about 10 cm thick.

Environmental Interpretation

Taieb and Tiercelin (1979, in press) have developed a generalized model of the Hadar Formation which is in the process of being refined with detailed micro-stratigraphic information (Tiercelin, in progress). They organized the paleoenvironmental setting at Hadar into three time stages. The focal element of their paleoenvironment is a major lake surrounded by marshy environments, and into which a few rivers flowed from the Ethiopian Escarpment. The rivers had developed flood plains, delta plains and deltas. During the first stage defined by Taieb and Tiercelin, represented by the Basal Member up through the Sidi Hakoma Submember #3 Sand, the deltaic and fluvial elements were prevalent at Hadar but lacustrine conditions existed just prior to deposition of the Sidi Hakoma Tuff. The second stage during the deposition of the uppermost Sidi Hakoma Member and lowermost Denen Dora Member was dominated by a full-scale transgression into the region by the lake. During the third stage of their model, deltaic and fluvial conditions returned to the site, beginning with the Denen Dora Submember #3 Sand. Fluvial conditions became well-developed by middle Kada Hadar and persisted through deposition of the formation.

Rather than focus the model of the environment on the lake, it is instructive here to refine the simple model presented by Taieb and Tiercelin, focusing on the fluvial elements of the paleoenvironment. The bulk of the sediment is not lacustrine but instead consists of sequences of fluvial channel deposits (sands) which make up about 20% of the formation and either overbank and/or interfluvial colluvium deposits (non-laminated silty clays), which make up 70% of the formation. Lacustrine or marsh deposits make up the remaining 10%.

The sands and silty clays occur in sequences suggestive of cyclical development. Typical units (which are the stratigraphic submember units) consist of a 5- to 8-m-dominant sand unit and 25 to 40 m of silty clay, with or without minor sands and/or laminated lacustrine clay.

The fluvial character of the major laterally extensive sands is well indicated by the scoured and channeled basal contact and the abundance of trough cross-bedding caused by scour and fill in the fluvial channel. The fluvial elements of the environment cannot be well specified, but probably involved a few braided medium-sized streams and inter-fluvial zones and possibly a major meandering river, with a broad flood plain near its terminus at a major lake. Our sedimentological interpretation is hampered by having worked on such a limited area of exposures. We do not know the range of sedimentary environments that existed over the much larger region around Hadar and how these environments could have shifted across the Hadar site during Hadar Formation time. We can generalize that near the top of the formation the Kada Hadar Member becomes more conglomeratic westward toward the Ethiopian Escarpment. Near the bottom of the formation the Sidi Hakoma Member becomes more lacustrine toward the east. For example, reconnaissance visits to Meschale, southeast of Ourda, indicate a dominance of laminated clays which probably correlate with the Sidi Hakoma Member. This small amount of regional data suggests the center of any major lake was to the east of Hadar, perhaps along the edge of the Galalu Plain (Fig. 1) during the early stages of uplift of the plain. As such the Hadar Formation would represent the gradual filling in of the lake by young volcanic detritus derived from the east along the rising Ethiopian Escarpment. We would project that much of the formation represents the distal deposits of rivers carrying this detritus in a delta-plain or alluvial-plain setting.

A major problem exists in ascertaining the nature of these rivers because the fluvial sediments in the Hadar Formation have characteristics common to both meandering and braided streams. The lack of a uniform fining-upward sequence within all of the sand units argues against their having originated in the laterally migrating channel of a major river which meandered across its flood plain, and is in harmony with a braided stream origin. Braided character of the streams is further supported by: (1) the occurrence of broad shallow cut-and-fill troughs in some sands; (2) reptile egg clutches emplaced in the middle of sand units rather than at the tops; (3) occurrence of planar cross-bedding that is suggestive of the transverse bars common in downstream portions of braided streams (Smith, 1970); and (4) the regular sheet-like shape of the major sand bodies. The argument against a meandering stream origin for the sands is tempered by observations of modern compound point bars of meandering rivers which allow considerable irregularity in the fining-upward trend (e.g., Harms et al., 1963).

The environment of deposition of the non-laminated silty clays that make up the bulk of a typical cyclical unit is uncertain. The lack of lamination and the occurrence of terrestrial vertebrate fossils argue for deposition on a land surface rather than a lake bottom. Evidence for paleosols, including calcrete formation in these units also supports a terrestrial surface of deposition.

A major difficulty is reconciling the large proportion of silty clay in the cyclical unit with a braided-stream origin for the associated sand. Braided streams are characterized by dominantly sand and gravel loads compared to a lesser amount of suspended silt and clay. Also such streams are usually confined in very broad relatively steep gradient channels with unstable sand banks. They are less prone to occupying the overbank region where silty clay could be deposited. The sheet sand which was described in the Lower Kada Hadar Member illustrates well the quandry in ascertaining the nature of the river that deposited it. This sand, which westward becomes a sheet gravel, has internal characteristics of a braided stream, yet it overlies and underlies a thick, non-laminated silty clay sequence. However, at the distal parts of such braided rivers in the delta plain region a meandering distributary pattern could develop with significant inter-fluvial or inter-distributary areas (as on the Ganges). In these distal areas, overbank flooding could occur and there the silty clay flushed from upstream could be deposited.

Alternatively, the non-laminated silty clay at Hadar may in part be overbank deposits of a major meandering river which was fed by braided tributaries. For such a meandering river to develop such thick silty clay sequences calls for a high rate of subsidence and a low rate of lateral migration of the river in order to accrete the deposits before they are removed by lateral erosion.

It is useful to examine two modern sedimentary settings in the Afar with elements common to the fluvial and lacustrine elements projected to have existed at Hadar (Figs.

5A and 5B). If some of the Hadar Formation sand units were deposited by a major meandering river, their thickness of 3 to 8 m would indicate a comparably deep river that was perhaps 10 to 20 times as wide (Allen, 1979). It would have compared in size to the present Awash and the sedimen-tological setting would be similar to the 40 km wide alluvial and delta plain now established by the Awash at Lake Gamari and associated lakes in the Tendaho Graben (Figs. 1 and 5A). An alternate modern sedimentological analog to the Pliocene paleo-setting at Hadar is the upstream stretch of the Awash River in the area of Lake Caddabasa, which is west of the Ayelu volcano and the town of Gawani (Figs. 1 and 5B). Lake Caddabasa is the remnant of a large Holocene lake which left significant diatomite deposits (Taieb, 1974). The former lake basin is mostly a swampy plain receiving sediment from numerous small braided streams coming from the escarpment on the east. The plain is subjected to a slight amount of dissection by the Awash and the small drainage ways leading from the plain to the Awash. The Awash presently flows at the eastern edge of the plain but it recently abandoned a meandering course through the center of the plain (Fig. 5B), as can be seen on the U.S. Defense Mapping Agency 1/250,000 map of the area. Thus the plain at Lake Caddabasa presents all three elements that have been suggested to account for the Hadar Formation: (1) distal braided rivers and their interfluvial plains and delta plains; (2) a major meandering river and its broad alluvial and delta plain; and (3) a lake with possible swampy borders and broad areas subjected to shallow flooding.

A point of importance at Hadar concerns the silty-clay flood-plain deposits and the evidence that at least an incip-ient calcrete-bearing soil profile developed on them inter-mittently during Hadar time. With highly seasonal runoff from the Ethiopian Plateau, one would project the delta-plain area that accumulated the dominantly silty clay Hadar Formation was subjected to annual overbank siltation, concom-itant with subsidence. However, the development of calcite nodules and partial calcrete zones implies that at times the flood-prone area that existed at Hadar was protected from sedimentation so that soil formation on exposed old sediment could outpace covering by fresh unweathered sediment (Allen, 1974). Also the "U"-shaped channels imply that parts of the Pliocene alluvial plain at Hadar ceased alluviation and underwent channelized erosion by relatively narrow channels 3 to 4 m deep.

An example of a complete paleo-calcrete was uncovered at the hominid-bearing Locality 200, excavated by Barbara Brown and A.E. Dole in 1974. The excavation produced no new

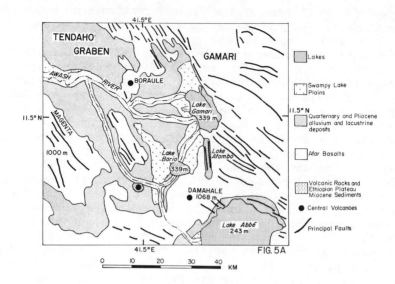

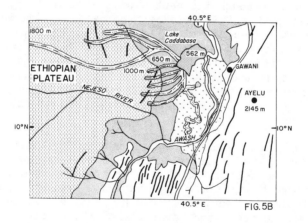

Figure 5.   A., B.   Two modern sedimentary environments along
the Awash River which serve as analogs of mosaic
environments that include elements which existed
at Hadar in the Pliocene.   Figure 5A shows the
terminus of the Awash River near Lake Gamari where
a low broad distal alluvial plain exists with
three principal distributaries and three lakes. In
Figure 5B, 120 km south of Hadar, the Awash River
flows to the side of a swampy plain that is the
remnant of a large Holocene lake that was fed by
the Awash River and several smaller streams coming
from the escarpment.   Both of these modern sedimentary basins have active faults on their margins
which probably project into the basins.

hominid or vertebrate material, but has proved to be very instructive geologically. A 3 m stratigraphic interval was excavated in the Sidi Hakoma Submember #1 about 5 m stratigraphically below the Sidi Hakoma Submember #2 Sand. This approximate stratigraphic interval has produced considerable hominid material such as at Locality 129 (Fig. 2), which also has a major calcrete. The excavation at Locality 200 revealed a 10- to 15-cm-thick siltstone that was thoroughly calcite-cemented. Extending 15 cm above this zone was a horizontal layer consisting of an anastomosed network of calcite root casts connecting irregular disc-shaped calcite nodules up to 30 cm in breadth. A blocky fractured argillaceous siltstone overlies the calcite zone and continues up to the Sidi Hakoma Submember #2 Sand. The calcite zone, the overlying fossil roots with a lateral orientation at the calcite-cemented zone, and the overlying blocky fractured siltstone together represent a mature calcrete-bearing soil profile. Such soils are thought to require periods on the order of a thousand to a few thousands of years to form (Williams and Polach, 1971), without the continual deposition of new sediment (Allen, 1974).

How could large portions of the flood-prone area of annually flooding major rivers have been interrupted from receiving sediment and, indeed, have been subjected to erosion for a long period of time? The answer may lie in the existence of faults active during deposition hypothesized to explain the marked eastward thickening of the Sidi Hakoma Member of the formation. Periodic small uplifting of the western block (absolute subsidence of the eastern block) on such faults would remove that portion of the flood plain from sediment accumulation. At that point soil development could proceed on the last flood deposits. In the early stages, under a moderately dry climate with a rainy season, calcite nodules would develop. With enough time free from new overbank sedimentation, a calcrete zone could develop. At the same time, erosion on small streams would proceed headward from the low scarp to create channels of the elevated flood plain. Detailed consequences of the fault hypothesis for explaining variation in sediment thickness, calcrete formation and channelization need to be checked in the field by locating the responsible faults and determining if the above features are associated with the appropriate fault block.

The modern analog sedimentological environments at Lake Caddabasa and at Lake Gamari (Figs. 5A and 5B) illustrate the interaction of faulting and deposition that is projected to have occurred at Hadar. At Lake Caddabasa several normal faults create escarpments in the volcanic rock highland south of the vast swamp plain. Probably some of these faults continue northward into this broad basin, but if any topographic expression ever existed it is now buried by sediment. Similarly some of the many block faults in the volcanic highland surrounding the alluvial and delta plain of the modern Awash River northwest of Lake Gamari (Fig. 5A) can be expected to continue into the plain. During faulting they would become topographically expressed and influence sedimentation, soil formation, and development of dissection channels.

Lacustrine-related sediments in the Hadar Formation provide major constraints to reconstructing the environment. Their existence within the fluvial sequence is consistent with the picture of the distal delta-plain area we projected above for either braided or meandering river systems near their base level at a lake. The gastropod/pelecypod layers occur in two ways, both of which are laterally extensive and probably of lacustrine origin. One has shells dispersed in silty clay; the other has thinner layers of very dense accumulations of shells. Freshwater gastropods and pelecypods can inhabit both lakes and rivers. High populations have been observed to develop very quickly in modern East Africa when a new, shallow, freshwater ponding situation occurs (McLachlon, 1974). Under such a situation primary productivity of plant food and detritus food for these molluscs is highest. The lake transgression at Hadar could have lasted for a relatively short time in the case of the dense layers, compared to the thicker mud sequences with dispersed gastropods. The dense layers which preserve a life assemblage indicate that some form of sudden death affected the population, such as by withdrawal of the water from the low plain.

The lignite represents a probable wetland marsh which presumably was formed by very shallow ponding of the flood plain and development of an emergent vegetation. The two marker clays, including the Triple Tuff clay with its abundant fish and ostracods, also seem best ascribed to either a lake transgression or to tectonic ponding of the river. A full transgressive sequence is indicated at Locality 147 (Fig. 2). First there was peri-lacustrine lignite deposition in an extensive marsh. This was followed by the development of shallow water algal mats which produced the stromatolite seam above the lignite. This was then followed by deposition of the two dense gastropod layers, then muds with dispersed gastropods and finally the Triple Tuff clay.

Summarizing, the sedimentological model for the Hadar Formation focuses on the delta plain of one or more rivers that flowed into a lake. The rivers had characteristics of braided streams. These rivers either had low interfluvial areas which vertically accreted significant amounts of silty clay or else they delivered their sediment to the alluvial and delta plain of a major meandering river. The basin subsided along with sedimentation during Hadar time, and faulting caused narrow blocks on the western part of the basin to be subjected to preferential development of soils and channel erosion and to less overall accumulation of sediment. The lake, which existed to the east of the low plain area at Hadar, shallowly flooded the region during late Basal Member time, late Sidi Hakoma Member time and in early Kada Hadar Member time. The middle transgression was the most significant.

Locality 333

As an example of the application of the sedimentological model it is instructive to look at the results of an excavation of locality 333. This locality has produced well over a hundred hominid fossil bones which can be shown to have derived from no less than 13 individuals, ranging in age from infant to adult (Johanson and White, 1979). The majority of the material has been found as float along an unusually high slope extending down from a cliff exposure of the resistant Denen Dora Submember #3 Sand in the Kada Hadar drainage (Fig. 2). Nearly all of the fossil material recovered from the slope is hominid; very few other species are represented. This in itself is remarkable considering how rare the hominids are and how prone their remains would have been to dilution by bones of the more common Hadar fauna.

A large portion of small post-cranial fossils were recovered from the slope, especially hand and foot bones. One articulated partial foot was recovered encased in a carbonate concretion that had broken across the foot bones. The slope fossil material, being exclusively hominid and from a life-assemblage of varying chronological age, suggested that a hominid social group was overcome, buried, and preserved by a Pliocene sedimentation-related event.

The locality was subjected to a careful and difficult excavation on the steep slope under the direction of Michael Bush during the last field season (1976-1977). The locality is divided into a north and south half separated by a low east-west ridge coming down from the resistant sandstone. Since the great majority of the hominid float was found on the slope north of this dividing ridge, the excavation began on the north slope. It consisted of a series of 3-meter-wide steps beginning 7 m below the prominent Denen Dora Submember

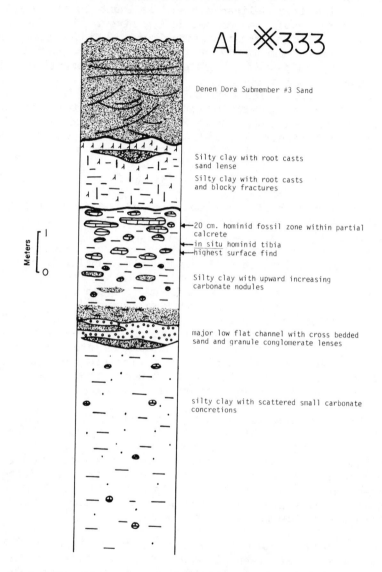

AL ✕333

Denen Dora Submember #3 Sand

Silty clay with root casts
sand lense

Silty clay with root casts
and blocky fractures

←20 cm. hominid fossil zone within partial
calcrete
←in situ hominid tibia
←highest surface find

Silty clay with upward increasing
carbonate nodules

major low flat channel with cross bedded
sand and granule conglomerate lenses

silty clay with scattered small carbonate
concretions

Meters

1

0

Figure 6.   Stratigraphic section at Locality 333 showing the
location of the hominid layer in the silty clay
sequence below the Denen Dora Submember #3 Sand.

#3 Sand and extending up to the sand. This sand is about 20 m above the Triple Tuff marker clay at this location and most of the section in between these two horizons is silty clay.

The excavation on the north slope provided an excellent geologic section. Three low flat isolated channel fills occur in the silty clay sequence. The low broad channels are about 15 m wide and 50 cm thick and contain cross-bedded sand and small lenses of granule conglomerate with occasional fish and rodent bones. Low down in the excavation the silty clay contains a few scattered small carbonate nodules. These increase in abundance and in size becoming disc-shaped about 2.5 m below the base of the prominent sand. Further upward the nodules continue to increase and merge. Irregular discontinuous solid carbonate lenses occur that are 10 cm thick and up to 2 m long. It is in this zone of dense carbonate where the hominid fossils were ultimately to be found. Just above this, at 2 m below the prominent sand, there is an abrupt transition of the nodule-rich silty clay to 1.5 m of nodule-free silty clay with a dense development of blocky fractures and vertical carbonate-filled root casts. This is followed up to the prominent sand by an irregular broad thin lens of fine cross-bedded sand overlain by a thin silty clay that is extremely rich in root casts. A scoured topography exists at the base of the Denen Dora Submember #3 Sand. The 5-m-thick sand consists of medium- to large-scale trough cross-beds in the lower portion and plane-bedded sand in the upper portion. Pieces of driftwood are abundant in the top of the sand north of this locality. Figure 6 shows the geologic section of the excavation.

No hominid material was found in the initial 3-m-wide step section. The upper 2.5-m section of the excavation was extended 2 m to the south to the dividing ridge. The first hominid material was recovered from the dense carbonate nodule zone near the dividing ridge. Most of the excavated fossils were found on extending the excavation of the hominid layer 5 m south of the dividing ridge. Of the 18 hominid fossils found in the excavation, 17 were narrowly confined to a 20 cm vertical interval that is about 40 cm above the highest slope find. The other specimen, a broken tibia, came from 40 cm lower than the main hominid layer. It was implanted vertically, upside-down with the upright broken distal end encapsulated and coverd by a network of small carbonate globular nodules and tubes.

The distribution of the hominid bones on the excavation floor was scattered in patches over a span of 7 m. The excavated fossils consisted of isolated teeth and post-cranial bones including two articulated finger bones. One individual clavicle bone was broken and separated by 40 cm. At least three individuals are represented, one an infant, one an immature adolescent and one an adult (Bush, Michael, and Johanson, D.C., pers. comm.).

All of the excavated material in the entire trench was carefully screened for fossils and yet very little non-hominid fossils were recovered. These consisted of a few unidentified bone fragments, two crocodile teeth, and several fish and rodent bones. Most of these were recovered from within the hominid stratigraphic interval. Some of the rodent bones were localized, possibly as the remains of owl pellets (William Kimbel, pers. comm.).

Thus the excavation produced fossils in place that match the abundant slope finds in terms of being dominantly hominid and of varying chronological age. The excavated fossils were focused stratigraphically and areally just above the highest slope find. It is straightforward to judge that the excavated concentration is the remnant of a larger concentration that had been eroding to produce the slope finds. Together, the distribution of slope and excavated fossils indicates the area of concentration was somewhat elongate in a north-northeast to south-southwest direction. Though most of the north end of the concentration appears to have eroded, more fossils could well exist within the unexcavated stratum at the south end.

The scattered, disarticulated, broken skeletal remains in the excavation suggest scavenging of the hominid bodies occurred on the depositional surface. However, the bones show virtually no evidence of gnawing marks that would have been left by large carnivores (Johanson, D.C., pers. comm.).

If the slope fossils and the excavated fossils indeed were derived from the same depositional interval, as seems likely, then this depositional surface was sedimentologically diverse, as indicated by the variation in the matrix of some of the slope finds. This matrix includes massive silty clay and carbonate nodule similar to that in the excavated hominid layer. However, one specimen, the skull of a baby, was recovered from a laminated fine-sand matrix and another had an accumulation of pea-sized carbonate balls of unknown origin as an adhering matrix.

The sedimentological model provides a framework into which an environmental reconstruction of Locality 333 can be fitted. We will start with the major lake transgression that produced the lignite, stromatolite, gastropod layers and Triple Tuff marker clay during late Sidi Hakoma time. The sequence includes about 30 m of mostly silty clay. Not far from Locality 333 an entire articulated elephant skeleton was recovered from silty clay below the Triple Tuff marker clay. The elephant may have become bogged down in the shallowly submerged area either travelling over a flooded delta plain or feeding on emergent vegetation in the lake. A low, perfectly flat depositional surface existed over the whole Hadar site when the Triple Tuffs and the associated Triple Tuff marker clay were laid down.

The Hadar area then emerged and resumed its former delta-plain status. Silty clays accumulated, and shallow channels with slab-like cross-sections were cut and filled with cross-bedded sands at Locality 333. These channels could have represented distributary channels on the delta plain or, crevasse channels which conducted flood waters from the major river channel onto the flood basin where mostly silty clays accumulated. A few fish bones and crocodile teeth were occasionally incorporated into the silty clay during flood deposition and scattered rodent bones accumulated at times when the plain was emergent.

During one of these flood events on the low alluviating plain, a large group of hominids was killed, in mass. The depositional surface was one where slow overbank flow was adjacent to faster currents able to sort and laminate fine sand. Among the many possibilities, the group could have either been directly drowned and buried with little scattering by currents, or the group could have become bogged down while trying to cross the wet plain. The group was only partially buried in the mud after the flood waters subsided. The carcasses were then subjected to minor scavenging so as to cause partial disarticulation and loss of skeletal material. The scavenging might have been done by crocodiles and fish, rather than by large carnivores. Hands and feet were preferentially preserved, possibly because these parts may have been thrust into the mud when the individuals fell.

Silty clays continued to accumulate to a thickness of about 1 m above the hominids, perhaps in several annual floods. A small channel developed and filled with sand. The plain then underwent a major long-term emergence. Dense grass developed and a partial calcrete began to develop in the soil profile. Carbonate accumulated around the hominid bones. The long-term emergence may have been caused by relative uplift along a fault of part of the plain as proposed above. Alluviation of the downthrown block and subsidence re-established a uniformly low area over which braided or meandering rivers moved across the Hadar site laying down the Denen Dora sheet sand.

Summary

The Hadar site in Pliocene time was within the sedimentary basin that developed in the Afar depression between the Ethiopian Plateau and the west-tilting central Afar fault blocks. The adjacent regions were volcanically active. The cyclical silty clay-sand volcaniclastic sequence at Hadar was mostly deposited in a low-lying delta or distal alluvial plain. Here silty clays accumulated as flood overbank deposits in interfluvial or interdistributary areas. The major sands are sheet-like in character with internal characteristics, including the lack of a uniform fining-upward sequence, that are more consistent with deposition by braided rivers rather than by meandering rivers. The large overall proportion of non-laminated silty clay in the Hadar Formation is indicative of the low-lying delta plain portion of the fluvial setting.

The inter-fluvial or delta-plain silty clay deposits were intermittently subjected to prolonged periods without sedimentation during which time they underwent soil formation and channeled erosion. Possibly this was due to block faults active during sedimentation which caused the sedimentary sequence to vary in thickness across the site.

A lake existed beyond the delta plain and underwent three principal shallow transgressions of the low Hadar area. The middle period, when the upper Sidi Hakoma Member and lower-most Denen Dora Member were deposited, was the most extensive.

Toward the end of Hadar time uplift of the Ethiopian Escarpment may have accelerated, and thereby resulted in the appearance of the abundant conglomerates in the Kada Hadar Member which increase westward.

The Hadar fossil vertebrate fauna is unusually rich. The fauna is particularly concentrated stratigraphically in the Sidi Hakoma and Denen Dora Members and geographically in the Kada Hadar and Sidi Hakoma tributary sectors. This could be due to a fortuitous combination of a productive ecological environment and climate in a low-lying delta-plain geologic setting where animals might have been trapped in overbank sediments and preserved in an articulated condition or have their remains transported and deposited in the river channel itself. The narrow geographic confinement of the abundant fossils may further be related to faults active during sedimentation which made certain fault blocks more predisposed in this regard than adjacent ones.

The detailed sedimentological study by Tiercelin (in progress) will undoubtedly shed considerably more light on the environmental setting at Hadar. It is important to recognize that relatively little time has been spent in field study at this exceedingly prolific site. Certainly more detailed field studies are needed at Hadar, such as were done at Locality 333, in order to interpret the paleoenvironment. Equally valuable will be the opportunity to explore a much larger region of the sedimentary basin in order to refine our view of the regional context of sedimentary environments that existed at the Afar margin in Pliocene time, and which could have shifted across the Hadar Site.

Acknowledgments

We are grateful to Don Johanson and Jean Jacques Tiercelin for several discussions pertinent to this paper. We also appreciate discussions with our colleagues, particularly Barbara Brown, Michael Bush, B.T. Gray, William Kimbel, R.C. Walter and T.D. White. Robert Sedivy provided point-counts of the Hadar sands.

We appreciate the opportunity offered by George Rapp, Jr., and Carl Vondra to present this paper in the AAAS Symposium.

Karen Toil deserves special credit for helping us prepare this paper under duress and for her cheerful disposition while doing so. This paper is Contribution No. 135 from the Department of Geological Sciences at Case Western Reserve University, Cleveland, Ohio.

References

Allen, J.R.L., 1974, Studies in fluviatile sedimentation: implications of pedogenic carbonate units, Lower Old Red Sandstone, Anglo-Welsh outcrop: Geol. J., 9, pt. 2, p. 181-208.

Allen, J.R.L., 1979, Studies in fluviatile sedimentation: an elementary geometrical model for the connectedness of avulsion-related channel sand bodies: Sed. Geol., 24, p. 253-267.

Aronson, J.L., Schmitt, T.J., Walter, R.C., Taieb, M., Tiercelin, J.J., Johanson, D.C., Naeser, C.W., and Nairn, A.E.M., 1977, New geochronologic and paleomagnetic data for the hominid-bearing Hadar Formation, Ethiopia: Nature, 267, p. 323-327.

Aronson, J.L., Walter, R.C., Taieb, M., and Naeser, C.W., New geochronological information for the Hadar Formation and the adjacent central Afar: Proc. VIII Pan African Cong. Prehist. and Quaternary Studies, in press.

Barberi, F., Borsi, S., Ferrara, G., Marinelli, G., Santacroce, R., Tazieff, H., and Varet, J., 1972, Evolution of the Danakil Depression (Afar, Ethiopia) in light of radiometric age determinations: Geol. J., 80, p. 720-729.

Beden, M., 1979, Les elephants (Loxodonta et Elephas) d'Afrique orientale: Doct. d'Etat. Fac. Sc. Poitiers, 2 vol., p. 567.

Cooke, H.B.S., 1978, Pliocene-Pleistocene Suidae from Hadar, Ethiopia: Kirtlandia, 29 (Cleveland Museum of Natural History).

Gray, B.T., Environmental reconstruction of the Hadar Formation, (Ph.D. thesis): Case Western Reserve University, Cleveland, in preparation.

Harms, J.C., Backenzie, D.B., and McCubbin, D.G., 1963, Stratification in modern sands of the Red River, Louisiana: Geol. J. 71, p. 566-580.

Harris, J.M., and White, T.D., 1977, Suid evolution and correlation of African hominid localities: Science, v. 198, p. 13-21.

Johanson, D.C., Taieb, M., Gray, B.T., and Coppens, Y., 1978, Geological framework of the Pliocene Hadar Formation (Afar, Ethiopia), in Bishop, W.W., ed., Geological Background to Fossil Man: Scottish Acad. Press, Edinburgh, p. 549-564.

Johanson, D.C. and White, T.D., 1979, A systematic assessment of early African hominids: Science 202, p. 321-330.

Leakey, M.D., Hay, R.L., Curtis, G.H., Drake, R.E., Jackes, M.K., and White, T.D., 1976, Fossil hominids from the Laetolil beds: Nature, 262, p. 460-466.

McLachlon, A.J., 1974, Development of some lake ecosystems in tropical Africa, with special reference to the invertebrates: Biol. Rev., 49, p. 365-397.

Roche, H. and Tiercelin, J.J., 1977, Decouverte d'un industrie lithique ancienne in situ dans la formation d'Hadar, Afar central, Ethiopie: C.R. Acad. Sci., (Paris) Series D, 284, p. 1871-1874.

Smith, N.D., 1970, The braided stream depositional environment: Comparison of the Platte River with some Silurian clastic rocks, North Central Appalachians: Bull. Geol. Soc. Am., 81, p. 2993-3014.

Taieb, M., 1974, Evolution Quaternaire de bassin de l'Awash, (thesis): Universite de Paris.

Taieb, M., Coppens, Y., Johanson, D.C., and Kalb, J., 1972, Depots sedimentaires et faunes du plio-pleistocene de la basse vallee de l'Awash: C.R. Acad. Sci. (Paris) Serie D, 275, pp. 819, 822.

Taieb, M., Johanson, D.C., Coppens, Y., and Aronson, J.L., 1976, Geological and Paleontological background of Hadar hominid site, Afar, Ethiopia: Nature, 260, p. 289-293.

Taieb, M. and Tiercelin, J.J., 1979, Sedimentation pliocene et paleoenvironments de rift, example de la formation a hominides d'Hadar (Afar, Ethiopie): Bull. Soc. Geol. de France, I, p. 243-253.

Taieb, M. and Tiercelin, J.J., La formation d'Hadar, depression de l'Afar, Ethiopie, stratigraphie, paleoenvironments sedimentaires: Proc. VIII Pan-African Cong. Prehist. and Quaternary Studies, in press.

Tiercelin, J.J., Sedimentation de rift: exemples de la actuel et de la formation pliocene d'Hadar (tentative title), (thesis): Doct. d'Etat, Fac. Sc. Marseilles, in preparation.

Walter, R.C., Volcanic rocks of the Hadar Formation and adjacent Afar areas: petrology, geochemistry and geochronology, (Ph.D. thesis): Case Western Reserve University, Cleveland, in preparation.

Williams, G.E. and Polach, H.A., 1971, Radiocarbon dating of arid zone calcareous paleosols: Bull. Geol. Soc. Am. 82, p. 3069.

Gary D. Johnson, Pamela H. Rey,
R. H. Ardrey, Charles F. Visser,
Neil D. Opdyke, R. A. Khan Tahirkheli

7. Paleoenvironments of the Siwalik Group, Pakistan and India

Introduction

The study of sedimentary successions is often characterized by an inability to establish or recognize isochronous surfaces. As a result, the rates of sedimentation and the style and frequency of recurring sedimentologic events such as soil forming processes and their duration are often difficult to reconstruct. This is particularly true of fluvial successions in which the cut and fill relationship generally associated with lateral accretion deposits (channel) results in the development of multiple hiatus' or diastems representing an unknown record, now eroded, of sedimentary events. In this context soils developed on aggrading alluvial landscapes may be preserved in the rock record (Ruellan, 1971; Gerasinov, 1971; Johnson, 1977). If sedimentation events are frequent, preservation of pedogenically young soils is to be expected. Root casts, crotovina, translocated clay and other pedological features may be preserved by burial and serve in later identification of a fossil soil (Teruggi and Andreis, 1971).

The sediments studied in this report are Late Miocene to Pliocene in age (circa 12 m.y. B.P. to 2.3 m.y. B.P.) and record climatic changes which are suggested by paleontological study. Analyses of mammalian fossil assemblages and pollen suggest that the climate was warm and humid during the deposition of the lower part of the sedimentary sequence and changed to a more temperate and seasonally drier climate during the upper part of the sequence (Nandi, 1975; Tattersall, 1969a and b; Leakey, 1969; and Prasad, 1971; Pilbeam, et al., 1977a, 1977b, 1979; Johnson, 1977; Moonen et al., 1978). If these climatic changes did take place, they should be reflected in associated fossil soils.

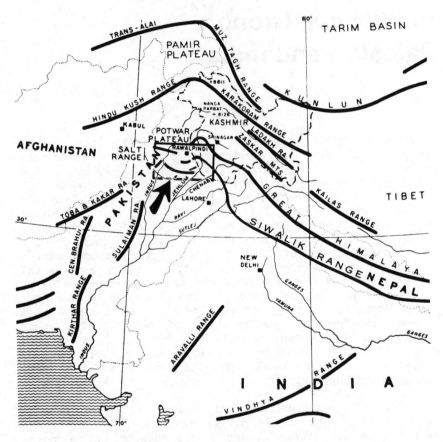

Figure 1. Location map of area in northern Pakistan.

With the advent of magnetic polarity stratigraphy (Opdyke, 1972) especially applied to terrestrial deposits (Johnson, N.M., et al., 1975), a means has become available to establish time lines in sedimentary deposits independent of radiometric dates. Recently, considerable new work has focused on establishing a magnetic reversal chronology for a number of important Late Tertiary and Quaternary terrestrial stratigraphic sequences in Africa, Asia, and North America which yield critical assemblages of fossil mammalian faunas. Emerging from these studies is an opportunity to observe variation in the character of sedimentologic events responsible for the style of deposition of the sediments which enclose the faunas.

The Late Tertiary and Quaternary deposits which lie to the south of the Himalaya Mountains in India and Pakistan present an opportunity to unravel the complex history of an evolving mountain range (Fig. 1). These sediments accumulated in a foredeep basin whose present surface expression is the Indo-Gangetic Plain and the adjacent faulted and folded foothills belt to the north of the plain. The cumulative sedimentary record of some 15 to 20 million years of uplift and denudation of the adjacent Himalaya is represented here. These sediments, the Siwalik Group and related rocks, also contain a varied record of terrestrial faunal and floral evolution during the Late Cenozoic and record the changes in sedimentary regime of fluvial systems which responded to continuing morphogenesis of this mountain belt. The magnetic polarity stratigraphy of the Siwalik Group has been variously reported, but the sedimentary, tectonic and geomorphologic implications of these data have not been fully developed (Keller, 1975; Keller, et al., 1977; Barndt, 1977; Kamei, 1977; Opdyke, et al., 1977a; Johnson, N.M., et al., 1977; Opdyke, et al., 1977b; Visser and Johnson, 1978; Opdyke, et al., 1979; Johnson, G.D., et al., 1979; Johnson, N.M., et al., 1980). This paper will summarize some of the salient features of fluvial sedimentation and associated pedogenesis (soil formation) in the Himalayan molasse in Northern Pakistan and India based on a chronology constrained by a paleomagnetic reversal stratigraphy of the Siwalik Group.

Paleoenvironmental Rationale

Several earlier studies of the clastic Siwalik Group have related the typical cyclic aspect of the multiple sandstone-mudstone couplets to varying, complex stream systems much like those developed today on the modern Indo-Gangetic Plain (Gansser, 1964). These stream sediments represent the product of aggrading fluvial sedimentation, and are characterized in part by fining-upwards sequences of

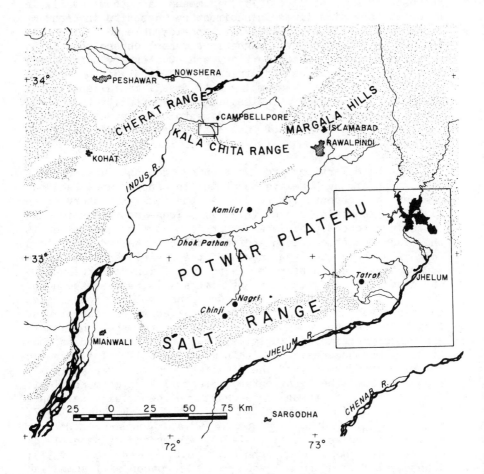

Figure 2. Potwar Plateau, Pakistan, showing location of
stratotypes of the Siwalik Group. The principle
sites in the Central Potwar discussed in text lie
between the villages of Dhok Pathan and Kamlial.
Other sites from the Eastern Salt Range discussed
in text are included in outlined area surrounding
the town of Jhelum (see Figure 3).

sandstones overlain by mudstones (Allen, 1965a, 1970). A genetic relationship has been demonstrated between these fluvial "cycles" and aggrading stream systems of varying sinuosity (Visser and Johnson, 1978; Halstead and Nanda, 1974). Very low sinuosity streams and braided stream systems (Miall, 1977) are indicated by certain Siwalik facies (Pilbeam, et al., 1979) and are generally important only in source-proximal sites within the sedimentary basin. The Siwalik Group is divided into many lithostratigraphic units based on the facies development of these fluvial cycles, a property closely related to variables of source area tectonics and basin aggradation.

Stratigraphic and areal studies of the Siwalik fluvial deposits have been difficult due to the lack of chronologic control for the highly variable, complexly interfingering facies encountered. Many of these problems are directly attributable to the multi-storied character of many of the lateral accretion (channel) sandstone and vertical accretion (overbank or flood basin) mudstone intervals which characterize the aforementioned sandstone-mudstone couplets. Most of the Siwalik sandstones that we have observed in the field are not single-story fluvial sands but are multi-storied sand bodies reflecting a complex fluvial history. Additionally, overbank deposits may not be single-storied either, but may reflect successive individual events. The present study examines the alluvial character of the Chinji through Upper Siwalik formations of the Siwalik Group in the central Potwar Plateau area of northern Pakistan and a portion of adjacent Kashmir (Fig. 2).

The Chinji Formation, exposed in the vicinity of the village of Khaur on the southern flank of the Khaur-Dhulian anticline in the central Potwar Plateau, contains mostly bright red mudstones interbedded with thin, single-storied sandstones. Approximately 300 m of Chinji is exposed (Cotter, 1933; Barndt, et al., 1978; Pilbeam, et al., 1977a, 1979; Johnson, N.M., et al., 1980). The Nagri Formation, exposed near the village of Dhok Pathan (Fig. 2), consists predominantly (80-85%) of thick, multi-storied sandstones interbedded with thin, dark red-to-orange mudstones. The sandstones are medium- to coarse-grained and frequently contain sheet gravels. Further to the east, near the village of Maluhwala, the mudstone to sandstone ratio increases in the Upper Nagri (Pilbeam, et al., 1979). Throughout the Nagri, the mudstones tend to change in color from dark red near the base of the formation upwards to a brighter orange color in the upper part of the formation. The section exposed between the village of Dhok Pathan and the Chinji-Nagri boundary on the Dhulian dome is approximately 1700 m thick.

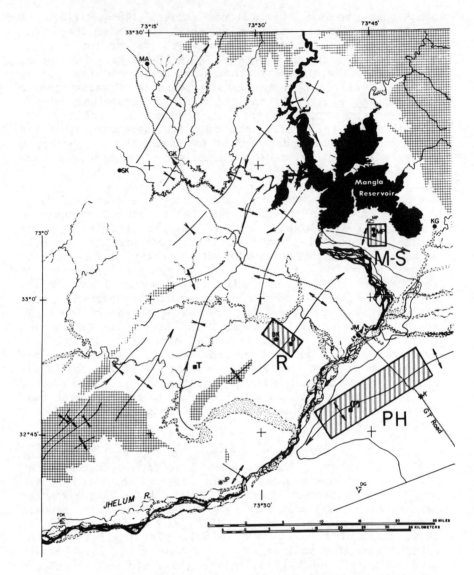

Figure 3. Outline map of the area in the vicinity of
Jhelum, Pakistan. Areas discussed in text as
follows: PH, Pabbi Hills anticline; R, Rhotas
anticline; M-S, Mangla-Samwal anticline. Cross-
hatched pattern represents areas having an elevation greater than 700 m. Important townsites
as follows: GK, Gujar Khan; JM, Jhelum; K,
Kharian; LM, Lala Musa.

The Dhok Pathan Formation near the village of Dhok Pathan, (Fig. 2) has deep red and brown-to-orange claystones interbedded with gray-to-white primarily single-storied sandstones (50% sandstone). Significant lateral facies changes are noted with sandstone bodies lensing in and out regularly. The Dhok Pathan is reported to be 320 m thick at Dhok Pathan village, but the top of the formation is not exposed.

The Siwalik Group as developed in southwestern Kashmir and in the hills of the Eastern Salt Range some 150 km southeast of the Dhok Pathan area was also studied (Fig. 3). Previous stratigraphic studies in the vicinity of the town of Jhelum, by Pilgrim (1913), Wadia (in Pascoe, 1930, and in Fermor, 1931), Gee (1945, 1947) and Wadia (1945), generally recognized the occurrence of Miocene, Pliocene and some Pleistocene clastic rocks exposed in the various structures making up the eastern extension of the Salt Range: the Chambal Ridge, the Jogi Tilla/Rhotas anticline, and the Kharian or Pabbi Hills anticline. Subsequently, work by de Terra and Teilhard de Chardin (1936) and the authors recognized the occurrence of Plio-Pleistocene rocks attributed to the Upper Siwalik Sub-Group in the Rhotas and Pabbi Hills structures, the north flank of the Jogi Tilla structure (Tatrot), and the Mangla-Samwal anticline near New Mirpur, in southwestern Kashmir (Fig. 3).

Fluvial Character of the Stratigraphic Succession

Outcrops of the Siwalik Group of northern Pakistan from the Eastern Salt Range and southwestern Kashmir illustrate the fluvial character of the group. Some 750 m of the Siwalik Group are exposed in the Mangla-Samwal anticline located just south of the town of New Mirpur in southwestern Kashmir (Figs. 3 and 4). This stratigraphic sequence records a system of stream-proximal clastic deposits dominated by point bar (lateral accretion) deposits with only minor development of overbank (vertical accretion) deposits. In general, the lower portion of the sequence is dominated by sediments deposited in sites proximal to the stream, while the upper portion is characterized by sediments deposited in more distal sites.

Constraining the age spectrum of this stratigraphic sequence are two bentonitized tuffs which straddle a zone of rocks representative of a transition from normal-to-reverse polarity magnetization (Fig. 4). The ages of the two ashes (Visser and Johnson, 1978; Johnson, N.M., et al., 1980) are respectively 2.5 and 2.3 $^+$ .4 m.y. The magnetic polarity transition which they straddle is the Gauss/Matuyama boundary

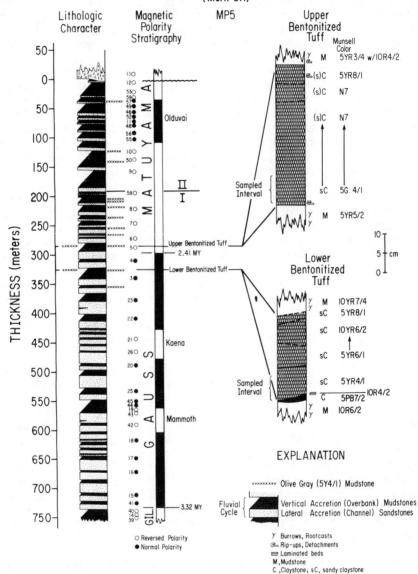

Figure 4a.   Sketch map of magnetic polarity zonation of
Siwalik sediments extrapolated from sites
determined along the Jhel Kas south of New
Mirpur, Kashmir. See Figure 4b for strati-
graphic context of this magnetic zonation.
Black refers to normal polarity.

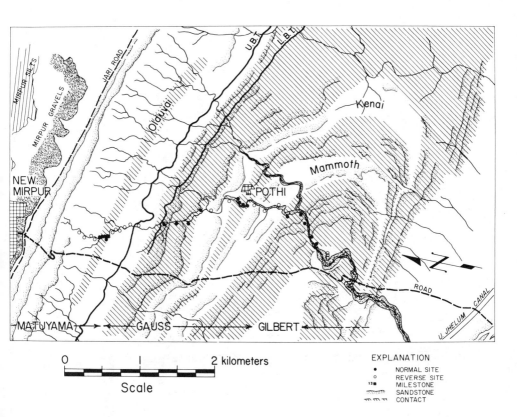

Figure 4b. Magnetic polarity stratigraphy of the Siwalik
section as exposed along the northern flank of
the Mangla-Samwal anticline in the vicinity of
New Mirpur, Kashmir. Schematic stratigraphic
column of lateral accretion sandstones (stippled)
and vertical accretion (overbank) mudstone units
are indicated. Units 1 and 2 in the stratigraphic section are referred to in text. From 3 to 5
oriented rock specimens were collected from each
of 59 sites located along the Jhel Kas, two kilometers south of New Mirpur. Details of associated volcanic ash couplet which straddles the
Gauss/Matuyama magnetic polarity transition are
detailed at right. For explanation as to tech-
nique and procedure of magnetic sampling program
see N. M. Johnson, et al. (1975) (From Johnson,
et al. (1979).

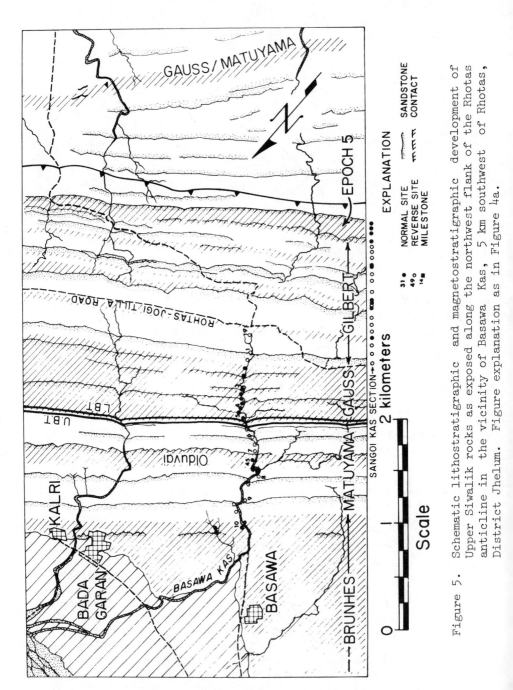

Figure 5. Schematic lithostratigraphic and magnetostratigraphic development of Upper Siwalik rocks as exposed along the northwest flank of the Rhotas anticline in the vicinity of Basawa Kas, 5 km southwest of Rhotas, District Jhelum. Figure explanation as in Figure 4a.

(Opdyke, et al., 1979). From this radiometric data the correlation of the magnetic polarity zonation is thus firmly established. The magnetic stratigraphy shows that the Mangla-Samwal section (Fig. 3) records a continuous series of sedimentary events from 3.3 m.y. B.P. to about 1.7 m.y. B.P.

The establishment of similar chronologies in the other structures of the southeastern Potwar and adjacent Kashmir area has made it possible to consider the overall effect of spatial relations within the Himalayan foredeep basin and the influence that position may have had on sedimentation during the Late Cenozoic (Visser and Johnson, 1978; Opdyke, et al., 1979; Johnson, G.D., et al., 1979).

Another long stratigraphic section in the Rhotas anticline (Fig. 5) may represent one of the longest continuous sedimentary sections thus far defined anywhere in the world in fluvial rocks (Fig. 6). Here a magnetic polarity section has been documented from the medial part of normal magnetic chron Five through the early Brunhes Normal Magnetic chron (Opdyke, et al., 1979); a record of some 5 million years, spanning the time from latest Miocene through to early Pleistocene (Fig. 6).

At approximately Olduvai time (earliest Pleistocene), a significant change in the sedimentary character of the Siwalik sediments occurred within the Rhotas region. Polymictic conglomerates of a general crystalline aspect were deposited at the 1450-m interval. Bedforms suggest a braided stream morphology and imply the onset of relatively high gradients for the stream system flowing through this region of the Himalayan foredeep. Older fluvial cycles in the Rhotas section are typified by fining-upwards sand bodies which may be characteristic of streams possessing low gradients and high sinuosity.

As in the Mangla-Samwal section, the magnetic record is uniquely identified by the presence of the two bentonitized tuffs mentioned above, which straddle the Gauss/Matuyama transition. A comparison of the two stratigraphic sections represented by the Jhel Kas section at Mangla-Samwal (Fig. 4) and the Sanghoi/Basawa Kas sections of the Rhotas anticline (Fig. 6) illustrates a general disposition of more stream-proximal deposits in the Mangla-Samwal structure than in the Rhotas structure. In general the greater occurrence of multi-storied sand bodies in the Mangla-Samwal area corresponds to its medial or axial position within the Jhelum structural re-entrant (see discussion of this feature in Visser and Johnson, 1978). Deposits in the Rhotas anti-cline are generally more stream distal and contain a higher

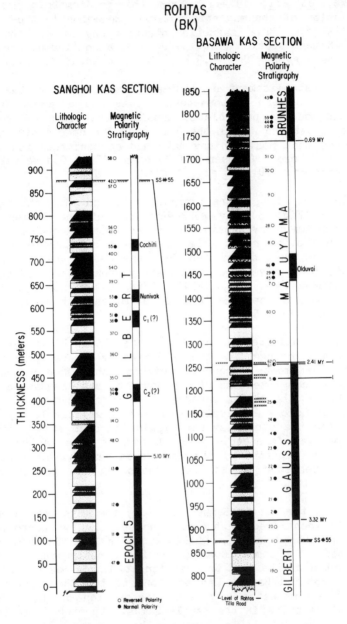

Figure 6.   Magnetic polarity stratigraphy of the Siwalik
section as exposed along the northwest flank of
the Rhotas anticline.   Data for Figure 5   is
represented in this figure.   Samples treated as
discussed in Figure 4b and 7.   From Johnson,
et al. (1979).

proportion of multi-storied overbank deposits containing numerous evidence of pedogenic (soil) modification.

Lateral Variation in Sedimentary Style

Isochronous events such as that of the bentonitized tuffs are one means for the discrimination of contemporary strata over a relatively large area. The lower and upper tuffs effectively confine an isochronous interval of fluvial sedimentation. Figure 7, a representation of lateral variability of fluvial cyclicity in eight outcrops of Upper Siwalik rocks in the vicinity of Jhelum, illustrates the problem of the lack of continuity of the deposits. The isochronous interval, constrained by volcanic ash, can be useful in describing the lateral changes in fluvial geometry within the sedimentary basin in existence at this time. At 2.41 m.y. B.P., sedimentation rates were nearly two times as great in areas associated with the apex and/or axis of the Jhelum structural re-entrant as reported by Visser and Johnson (1978).

Magnetic transitions recorded in sediments provided another means to trace the lateral persistence of lithologic units. In the Pabbi Hills anticline just south of Jhelum (Fig. 3), three stratigraphic sections spaced along a 14 km lateral traverse, transect the Olduvai normal magnetic event (1.86 to 1.71 m.y. B.P.). These sections (Figs. 8, 9 and 10) show the systematic migration between the position of the Olduvai event with reference to one of the sandstone units. On a gross scale the sandstones are qualified isochronous units but are reasonably valid as time indicators.

Assuming the duration of the Olduvai event to be constant in all localities of the Pabbi Hills, a 36-m difference in apparent depositional elevation is seen along the 14-km traverse during the 150-thousand-year period. The sedimentation rate at this time was approximately 41 cm per thousand years; the sedimentary subsidence differentiated along the 14-km traverse was 2.5 meters per km (Fig. 11). The general implication is that the net direction of thalweg migration during Olduvai time was westward along the line of traverse with an average rate of lateral migration of some 0.17 km/1000 years (Fig. 11).

The total thickness of lateral accretion deposits in a given section is influenced by many environmental variables. The thickness of an individual cross-stratified lateral accretion cycle, neglecting truncation and erosion, approximates the depth of the associated stream (Moody-Stuart, 1966; Allen, 1965a). The number of cycles observed in single- or

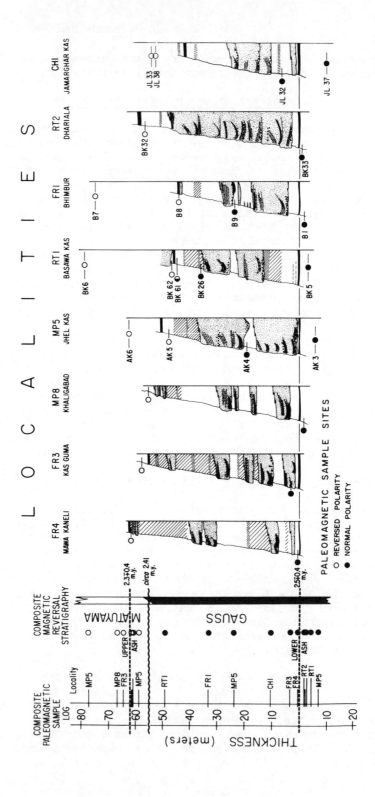

Figure 7. Lithostratigraphic variation at eight localities in the Jhelum area illustrating variation in Upper Siwalik sediments which are bounded by the volcanic ash couplet discussed in text. Paleomagnetic sampling at each site was in triplicate following procedures outlined by N.M. Johnson, et al. (1975). Reliability discussed in G.D. Johnson, et al., in preparation. Magnetic determinations by N. D. Opdyke. Data from the eight stratigraphic sections are projected onto a composite paleomagnetic sample log (left) from which a composite magnetic reversal stratigraphic column in the Jhelum area is erected. This shows the projected position at the Gauss/Matuyama polarity transition relative to the bentonitized tuffs of this report. Lithologic symbols: Lateral accretion sandstone facies, stippled pattern showing dominant bed form; vertical accretion mudstone facies with deposition fabric, no pattern; vertical accretion mudstone facies with evidence of pedogenic (soil) modification, hatched pattern. Inferred upper solum positions in pedogenically modified mudstones-A or B¹ horizons-, doubled hatched pattern. Modified from Visser and Johnson (1978). Reprinted by permission of the Geologische Rundschau.

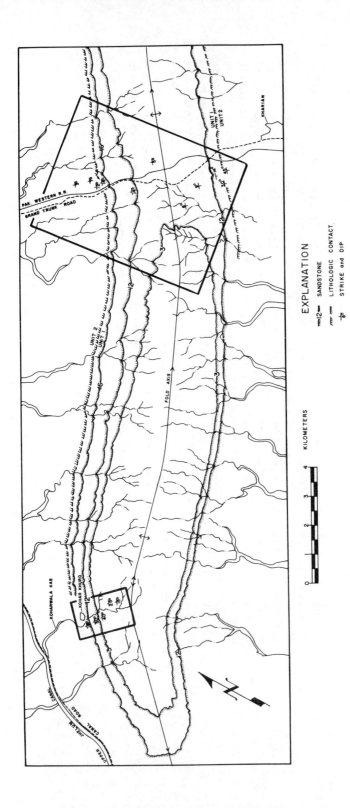

Figure 8. Outline map of principal lithostratigraphic units useful in differentiating structure of the Pabbi Hills anticline, northern Pakistan. Area of section 15, this report, outlined on left. Area of sections 16 and 2, of this report, is outlined on right. Modified from Keller, et al. (1977).

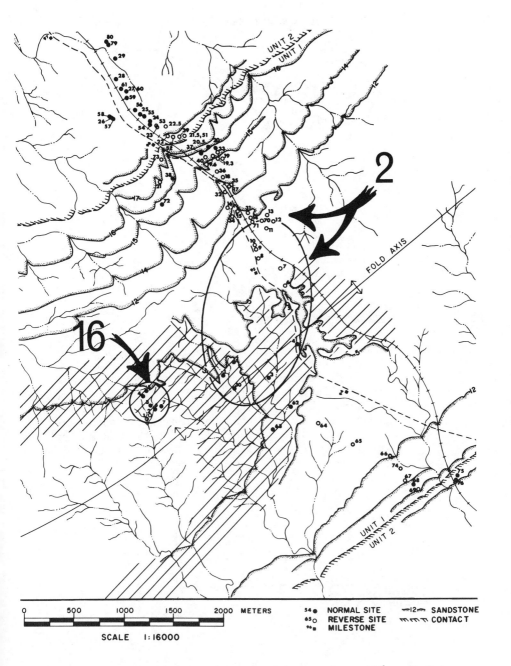

Figure 9.   Location of stratigraphic sections 16 and 2 in
            the vicinity of the Grand Trunk Road where it
            traverses the Pabbi Hills anticline southeast of
            Jhelum. Distribution of rocks known to be of
            Olduvai age represented by cross-hatched pattern.
            This zone represents Plio/Pleistocene boundary
            as seen in the Pabbi Hills. Data from   Keller
            (1975)   and   Keller   et   al.   (1977)
            Modified from Keller et al. (1977)

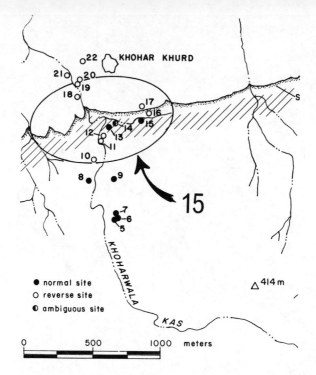

Figure 10. Location of section 15 in the vicinity of Khohar Khurd, western end of the Pabbi Hills anticline, south of Jhelum. Distribution of rocks known to be of Olduvai age represented by cross-hatched pattern. This zone represents the Plio-Pleistocene boundary in this portion of the structure. See Figure 8. Data from Keller (1975) and Keller, et al.(1977).Modified from Keller, et al.(1977).

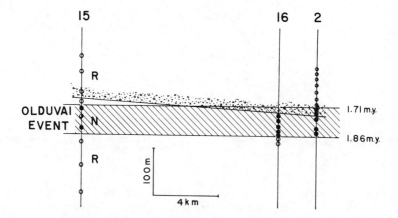

Figure 11. The nature of the time-transgressive aspect of Sandstone 3 (Ss 3), a principal lithostratigraphic unit of the Pabbi Hills, relative to the Olduvai normal event (see Figures 8 and 9).

multi-storied intervals may reflect the relative frequency of stream migration (Johnson and Vondra, 1972; Visser and Johnson, 1978; Leeder, 1975, 1978; Bridge and Leeder, 1979). Multiple encounters by a single stream over a particular site may imply oscillatory lateral migration with the rate of oscillation related to the rate of bank erosion and to the degree to which the overbank is confined to a restricted meander belt. Or, as an alternative model, multiple encounters may represent a thalweg which is free to sweep the flood plain. In both cases these variables are presumably related to variations in stream gradients and sinuosity. The multi-lateral encounters of more than one stream system as in anastomosing streams further complicates the system. Much of the Himalayan piedmont apron illustrates coalescing distributory systems. In spite of the many related variables, the total thickness of lateral accretion sandstone deposits at a site appears to be a valid measure of net proximal stream activity. In addition, the frequency of stream migration may also indicate proximity to major fluvial courses or axes. Localities near major fluvial axes during much of the interval of deposition are likely to display greater total lateral accretion thicknesses than distal flood basin localities. These general aspects can be seen in sections of the Pabbi, Rhotas and Mangla-Samwal structures and in the Jhelum area as a whole (Fig. 12).

Fluvial Cycle Recurrence Rate

As indicated earlier, the nearness of a point to a major stream axis may be gauged by the fluvial cycle recurrence rate (FCRR). Duration of sedimentologic events interpolated from the above stratigraphic sections which are constrained by a paleomagnetic reversal chronology allow for the recognition of individual fluvial cycles which may represent 25- to 65-thousand-year events. A case in point may be the fining-upwards, single-storied cycles occurring in the Jhel Kas section of the Mangla-Samwal anticline (Fig. 4) which lies between the Gauss/Matuyama transition at 2.41 m.y. B.P. and the Matuyama/Olduvai transition at 1.86 m.y. B.P. This record of 550 thousand years contains fourteen single-storied fluvial cycles in the succession—a duration of forty thousand years per apparent sedimentologic event. In the Rhotas section, the FCRR is 37 thousand years for the same interval. In virtually all cases, the overbank deposits (floodbasin mudstones) contain evidence of multi-storied pedogenic (soil) modification and at various levels there are pedogenic horizons. The floodbasin sediments which lie in stream-distal positions record successive events of overbank flooding, but do not record successive events of thalweg encounter.

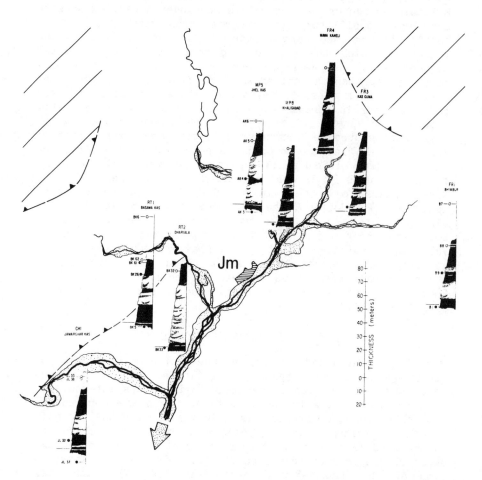

Figure 12.   Schematic   representation of fluvial facies   in
the Jhelum area plotted to illustrate   evidence
of a locus of prior stream activity is confined
to the axial region of the Jhelum structural
re-entrant. See Visser and   Johnson (1978)
for discussion of this structural feature.   The
intervals illustrated are confined by the   two
volcanic ashes mentioned in text.

Development of the
Floodbasin Stratigraphy

The development of the flood plain stratigraphy in Upper Siwalik sediments has been covered earlier by Visser and Johnson (1978). That discussion is reviewed here. Visser and Johnson (1978) point out that as the stream thalweg migrates laterally, relict channel and point bar deposits are covered by mudstones of the trailing flood plain. Repeated lateral migration and aggradation in this manner result in a sequence of fining-upward cycles. Thus, at any point in the fluvial plain, the vertical sequence of deposits gives a time record of the various fluvial environments crossing the point during stream migration. Erosion commonly takes place at the base of a cycle, rendering the previous cycle incomplete. Elimination of vertical accretion mudstones by this process creates "multi-storied" sand units. The study of various overbank sequences in the Jhelum area allowed for the recognition of several categories or types of vertical accretion (overbank) deposits on the basis of field identifiable characteristics (color, grain size, and internal fabric) (Visser and Johnson, 1978). These categories are as follows:

Type R1)　pale red, laminated claystones and silty claystones with fine lamination showing varying degrees of soft-sediment deformation, sparse crotovina (animal burrows) and very sparse plant debris; principal Munsell colors 5R6/2 to 5R4/2; lamination is accentuated by fine color variegation and parting, interbedded with brown siltstones (Munsell colors 10YR (5.5-6)/(2-3) of approximate 5 to 10 cm thickness.

Type Rm)　pale red, massive claystones and silty claystones with homogenized to massive fabric; Munsell color 5R6/2 to 5R4/2.

Type Yb)　variegated yellowish-brown claystones, silty claystones and clayey siltstones with homogenized to massive fabric; variegated color horizonation; principal Munsell colors 10YR (5-6) (4-6).

Type O)　olive-gray, olive-brown and dark yellowish-brown pisolitic claystones with homogenized to massive fabric; principal Munsell colors 5Y (4-5)/(1-4) and 10YR (4-5)/(2-4), black-to-brown pisoliths (3 mm diameter); transitional contacts with type Yb, olive or dark yellowish-brown horizons usually less than one m in thickness.

Type Sp) yellow, gray and brown sandstones and silty sandstones (splay deposits) with massive fabric; fine- to medium-grain size, fining-upward, poorly sorted; principal Munsell colors 5Y (6-7)/(1-4), N (5-6), 10YR (4-6)/ (2-3), 5YR (4-5)/(1-4).

The absence of extensive bioturbation in the laminated (type Rl) claystones suggests backswamp deposition with frequent ponding. Homogenized clays of the same color (type Rm) may represent bioturbated sediments of a more "ripened" aspect (see discussion of soil ripening in Pons and Zonneveld, 1965).

The homogenized fabric, variegated color horizonation, and the presence of sesquioxide pisolites in claystones of type Yb and type O represent evidence of pedogenic alteration of the original depositional fabric. Dark pisolitic zones in particular may be characteristic of illuviation in the B-horizons of these alluvial soils of Al and Fe sesquioxides (Al_2O_3, Fe_2O_3), clays, and some organic matter (Johnson, 1977). Dark green and olive-hued overbank sediments suggest reduced conditions typical of poorly drained soils (Papadakis, 1969).

Visser and Johnson (1978) applied a Markov chain analysis to a portion of the stratigraphic record limited to the Gauss/Matuyama polarity transition for a number of Jhelum area sites. Analyses of other ancient fluvial sequences have shown Markov chain analysis to be an effective means of detecting repetitive processes in multicomponent cyclic deposition (Gingerich, 1969; Allen, 1970; Selley, 1970; Miall, 1974). Applied to an array of thin, time-synchronous sections such as the present data base, the procedure derived the principal depositional facies present in a region of contemporaneous fluvial deposition. Since the depositional sequence is a result of laterally migrating fluvial environments, the areal distribution of facies in the flood plain can be ascertained.

Applying the method to the present stratigraphy, Visser and Johnson (1978) considered as depositional components the five aforementioned vertical accretion lithologies and a single variable for lateral accretion. Consideration of their data yielded the following:

1) Lateral accretion (La) tends to be followed by vertical accretion of type Yb. The fabric of type Yb appears to

be a product of pedogenesis (see 2, below). Type Rl also tends to follow lateral accretion, but with a somewhat lower probability than type Yb.

2) Type Yb tends to be followed by type O. On the basis fabric, color horizonation, and the presence of sesquioxide segregations it appears that type Yb/type O couplets represent soil horizonation. There is a slight tendency for type Yb and type O to alternate and in turn be followed by type Rm. The several instances of multi-storied type Yb/type O couplets appear to reflect successive episodes of soil horizon development which took place during the increment accumulation of overbank sediment. In this case type Rm occurs as a well-homogenized upper soil solum A-horizon with the underlying type O developing as an illuviated B-horizon.

3) Types Rm, Rl, and Sp tend to alternate with each other. The alternation of these vertical accretion lithologies suggests a multi-storied process of successive flooding, coarse slackwater deposition (type Sp), fine slackwater deposition (type Rl) and bioturbation (type Rm). Appropriately, splay sands (type Sp), tend to be followed by a laminated fine-grained lithology (type Rl).

4) Type Rl tends to be followed by lateral accretion (type La). If pale red laminated clays are the product of slackwater deposition in areas frequently flooded, the data support the tenable conclusion that laterally migrating stream thalwegs are immediately preceded by a frequently flooded proximal environment in which sedimentation overshadows pedogenesis. Marked pedogenic horizonation and disturbed vertical accretion fabrics usually give way to laminated claystones at some distance below the base of lateral accretion units. The exceptional cases, where lateral accretion sandstones overlie pedogenic vertical accretion units, can be explained by basal reworking of laminated topstrata.

Lateral accretion sandstones thus tend to be preceded stratigraphically by undisturbed vertical accretion units indicating a frequently flooded, sedimentation-dominant, flood basin environment. It was observed previously that lateral accretion units tend to be followed by vertical accretion units of type Yb, suggesting a post-migration environment of pedogenesis. Whether the environments preceding and following a laterally migrating thalweg indeed differ is uncertain; pedogenically modified vertical accretion lithologies overlying lateral accretion sandstones may reflect deep pedogenic overprinting occurring at a time much

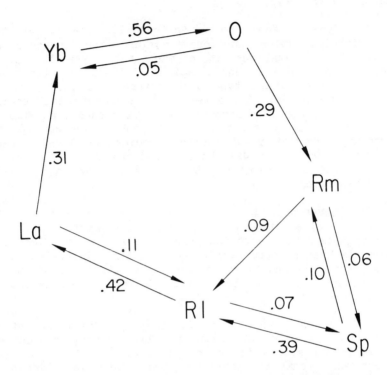

Figure 13.   Facies relationship diagram showing the transitions occurring more often than expected from a
random arrangement.   Paths of d 0.5 excluded for
clarity.   Values give the greater-than-random
probabilities of transitions.   Arrows point to
overyling unit.   Facies scheme as follows: La,
lateral accretion sandstones;   R1, pale red laminated claystones and silty claystones;   Rm, pale
red massive claystones and silty claystones;   Yb,
variegated yellowish-brown claystones, etc.;   0,
olive-gray, etc. pisolitic claystones;   and Sp,
splay deposits.   From Visser and Johnson (1978).
Reprinted by   permission of the   Geologische
Rundschau.

later than the active migration of the stream course (Visser and Johnson, 1978).

These relationships are summarized in Figure 13.

Evidence of Pedogenesis Affecting Floodbasin Stratigraphy

Although field criteria strongly suggest pedogenic modification of some of the above mentioned flood basin sediments, those lying below the upper volcanic ash (for example in the Jhel Kas and Rhotas sections) were analyzed further. The profiles were samples every 15 cm to a depth of 300 cm and were described in terms of textural units, contacts, color, bedding features and secondary features such as kankars and mottles.

Chemical analyses were conducted on eight profiles which were selected for kankar and iron concretion development and, to some extent, reddish coloration (i.e., 5 YR 4/4, 10 YR 5/4, 10 YR 7/4 on the Munsell color chart). Samples were ground and prepared for analyses according to Holmgren's (1967) sodium-citrate-dithionite extraction procedure. The extractions were then analyzed for extractable Fe, Al, Mn, and Si with a Perkin-Elmer Model 503 atomic absorption spectrophotometer.

Despite the problems presented by the study of paleosols developed on alluvium, many features reflecting progressive pedogenesis from ripening through gleying may be preserved in alluvial soils. Field and thin section observations point out evidence of soil ripening and subsequent formation of secondary features such as mottles, kankars, and iron concretions.

The first stage of pedological maturation, or "ripening" (Pons and Zonneveld, 1965), may be chemical, physical, or biological. Crotovina observed in these samples were regarded as evidence of biological ripening which results in homogenization of the soil substrate.

Signs of more advanced pedogenesis include iron-manganese concretions, mottles and kankars. Iron-manganese concretions are common in sedimentary rocks, surface deposits, or ancient soils, but their origin is obscure (Hunt, 1972). One variety of mottles consists of compounds of yellow-brown, brown or red color that result from a fluctuating water table and concomitant varying conditions of reduction (Birkeland, 1974). Moderately well-drained soils have mottles at depths of 2 to 3 feet, while imperfectly or somewhat poorly drained soils have mottles 6 to 16 inches below the surface (Hunt, 1972). Other light green or blue (5 G8/1 - 5 B7/1, Munsell color chart) mottles are here denoted "zones of reduction" and may indicate chemical reduction around decaying organic matter. Kankars form in calcareous parent material at depths related to the amount of precipitation received by the soil (Birkeland, 1974). In general, arid soils will have kankars within 1 m of the surface, while more humid climatic regimes, where rainfall exceeds 75 cm per year, will have kankars at depths greater than 1 m (Jenny, 1941; Sehgal and Stoops, 1972).

Holmgren's (1967) sodium-citrate-dithionite extraction method was followed to determine the concentrations of free oxides at regular intervals in the upper 300 cm of a soil profile. Sodium-citrate-dithionite extracts a certain proportion of amorphous and crystalline iron oxides and other anhydrous and hydrous oxides from soils (McKeague, et al., 1971). These oxide concentrations are compared to each other in order to make interpretations concerning horizonation within one profile, or the relative degree of pedogenic alteration between profiles formed on the same parent material (Asamosa, 1973).

Johnson (1977), Gile and Hawley (1966) and Juo, et al. (1974) have observed that free iron and/or Fe_2O_3 are positively correlated with clay in both primary depositional and secondary, pedogenic deposits. Since a fining-upwards sequence would have the Fe_2O_3 maxima close to the surface, and pedogenesis would result in downward migration of clays and Fe_2O_3, profiles of alluvial cycles which show Fe_2O_3 maxima below the finest-grained stratum indicate pedogenic evolution. Al_2O_3 also mirrors clay distribution (Buursink, 1971), and may be evidence of pedogenesis. Di-MnO maxima have been observed (via Holmgren's sodium-citrate-dithionite extraction method) lower in soil profiles than either di-Fe_2O_3, di-Al_2O_3 maxima or Mn oxide (Johnson, 1977; Blume and Schwertmann, 1969). The presence of Mn-rich concretions nodules and ped surface coatings usually indicates seasonal saturation with H_2O during soil formation (Soil Survey Staff, 1951), and in fact has been correlated with poor drainage conditions in a reducing environment (Blume and Schwertman, 1969). It would therefore be expected that high MnO values would be positively correlated with mottling and negatively correlated with values of di-Fe_2O_3, which are favored by oxidizing conditions. Finally, di-SiO_2 maxima may indicate buried land surfaces. Certain plants produce opaline silica phytoliths (Smithson, 1956) which may accumulate and indicate upper sola positions in the soil profile if the plants are buried and only the opal phytoliths are preserved

(Johnson, 1977). As already noted, however, A-horizons in which these elements would be concentrated tend to be obliterated in the soil record.

Horizonation is accompanied by color changes in parent materials which characterize various horizons. Grayish colors bluer than 10Y indicate gleying, while yellow-brown-to-red colors resulting from the presence of iron oxides and hydroxides are characteristic of B- and C-horizons (Birkeland, 1974).

Of eight sections analyzed, two were stream distal, four were transitional and two were proximal. Figure 14 is a sample description of a profile which incorporates field, chemical and thin section observations.

The integration of mobile oxide distribution of chemically analyzed profiles, thin section studies and field relationships reveal that immature mineral soils with kankars and iron concretions developed on the alluvial landscape of late Pliocene age. Although carbon analysis was not conducted, all horizons of these soils certainly contained less than 20 to 35 percent organic matter by weight and the soils were therefore classified as mineral soils (Soil Survey Staff, 1975). Young calcic horizons have 1 mµ accumulations which are soft and disseminated (Steila, 1976). Since the calcite localizations have not yet formed discrete horizons but appear disseminated throughout most horizons, these soils are considered immature. The preservation of ripening features in most profiles supports this conclusion.

Interpretation of Upper Siwalik Paleosols

The Siwalik mineral paleosols offer three parameters for general soil classification: (a) slight horizonation, indicating immature soils, (b) the occurrence of kankars throughout most profiles, and (c) a generally low organic carbon content. Kankars appear to occur stratigraphically close to inferred sola positions, although at times one pedogenic or fining-upwards unit may develop kankars due to pedogenesis in the overlying unit. Gile and Hawley (1966) note that illuviation through overlying sediments can affect paleosol evidence. The volcanic ash in profile KK-4 interrupts pedogenesis but probably does not remove sediment. This proximity to the soil surface of kankar development indicates that the soils are probably formed under a semiarid-to-arid climatic regime.

By analogy to modern soils developed elsewhere, these Upper Siwalik paleosols appear to be aridisols. With further

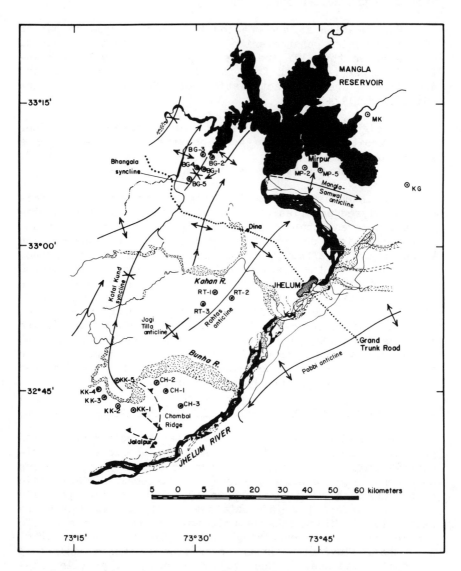

Figure 14a.   Location of Upper Siwalik paleosol sites
studied in the Jhelum area.

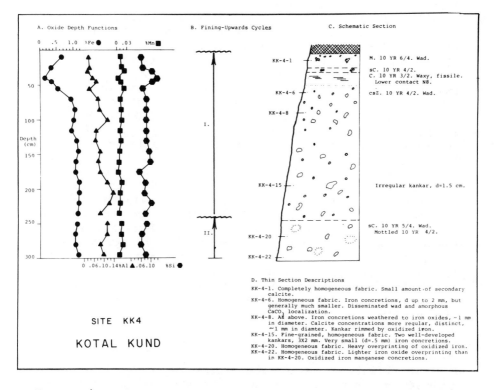

A. Oxide Depth Functions

B. Fining-Upwards Cycles

C. Schematic Section

KK-4-1 M. 10 YR 6/4. Wad.
sC. 10 YR 4/2.
C. 10 YR 3/2. Waxy, fissile.
Lower contact N8.

KK-4-6 csZ. 10 YR 4/2. Wad.

KK-4-8

KK-4-15 Irregular kankar, d=1.5 cm.

KK-4-20 sC. 10 YR 5/4. Wad.
Mottled 10 YR 4/2.

KK-4-22

SITE KK4

KOTAL KUND

D. Thin Section Descriptions

KK-4-1. Completely homogeneous fabric. Small amount of secondary calcite.
KK-4-6. Homogeneous fabric. Iron concretions, d up to 2 mm, but generally much smaller. Disseminated wad and amorphous CaCO$_3$ localization.
KK-4-8. As above. Iron concretions weathered to iron oxides, ~1 mm in diameter. Calcite concentrations more regular, distinct, ~1 mm in diamter. Kankar rimmed by oxidized iron.
KK-4-15. Fine-grained, homogeneous fabric. Two well-developed kankars, 3X2 mm. Very small (d=.5 mm) iron concretions.
KK-4-20. Homogeneous fabric. Heavy overprinting of oxidized iron.
KK-4-22. Homogeneous fabric. Lighter iron oxide overprinting than in KK-4-20. Oxidized iron manganese concretions.

Figure 14b. Schematic representation of typical paleosol development at site KK4, from the Kotal Kund region, 70 km southwest of Jhelum. Na-dithionite-citrate-extractable oxide distribution within the upper 300 cm of profile development capped by the upper volcanic ash discussed in text.

refinements in the definition of these soils (Soil Survey Staff, 1975) they may be interpreted as semiarid alfisols.

Sehgal, et al. (1968) describe aridisols occurring in the Punjab of India. Like the Siwalik paleosols, kankar development increases and organic matter decreases with depth, and colors range from dark yellowish-brown to moderate yellowish-brown. Brown (desert) soils, sierozems or aridisols are developing in India under a desertic, arid-to-semiarid or steppe-type climate with a mean annual temperature of 24.8°C, and mean summer and winter temperatures of 33.1°C and 14.9°C, respectively, and a mean annual precipitation of less than 500 mm. Similar soils are described by Sidhu and Gilkes (1977) in the semiarid climatic regime of the Indo-Gangetic Plain, where 70% of the 500 mm annual precipitation is received during July, August and September. The limited amount of water available in aridisols results in a lower intensity of chemical and physical reactions than observed in humid regions. As a consequence, the soils inherit the morphology of the parent material (Buol, et al. 1973), as observed in the preservation of color and texture in the Siwalik paleosols.

On the basis of nodular units of carbonate in paleosols developed in the vertical accretion deposits in the Old Red Sandstone of Britain, Allen (1974) hypothesized that Old Red paleosols formed in a warm-to-hot, semiarid-to-arid climate marked by seasonal rainfall. He suggested that the mean annual temperature was 16° to 20°C and the annual precipitation 100 to 500 mm. Deserts in New Mexico (Gile, 1960; Gile and Hawley, 1966) exhibit well-developed horizons of calcite and poor horizonation. Since the Siwalik soils exhibit a dispersed calcite "horizon" and distinguishable horizonation, an arid desert climate such as that defined by Gile is probably drier than the Siwalik paleoclimate.

By definition, aridisols experience very little clay illuviation and brief, transient influence from surface water (Soil Survey Staff, 1975; Steila, 1976). Conversely, alfisols may occur in seasonally arid moisture realms and may show signs of wetness, such as mottles and iron-manganese concretions, calcite localization and clay illuviation (Steila, 1976). The Upper Siwalik paleosols possess these characteristics of alfisols. Since calcification seems to be the dominant pedogenetic process, these soils are best described as semiarid alfisols.

Hunt (1972) described vegetation over desert soils to be dominated by shrubs, occasional trees and some woodland. The vegetation over aridisols consists of ephemeral grasses, and scattered xerophytic plants, although vegetation might not exist without irrigation (Soil Survey Staff, 1975). Semi-arid alfisols in the southwestern United States supported a vegetation of annual grasses and woody shrubs prior to culti-vation (Soil Survey Staff, 1975). The greenish zones of reduction observed in the Upper Siwalik paleosols suggest that some vegetation did exist, but the overall lack of microflora and phytoliths in these sediments could reflect a sparseness of vegetation above soils developed in a semiarid temperature and moisture regime as described above.

Studies of soils developed since the late Pleistocene on the Punjab plains give an estimate of the time required for semiarid soils to form. Ahmad, et al. (1977) found secondary accumulations of calcium carbonate in the subsurface of 10,000- to 20,000-year-old soils, while 2300-year-old soils had none. The older soils also exhibited clay illuviation and low organic matter content. Allen's (1974) study of Old Red paleosols concluded that 10,000 to 30,000 years are required to form calcrete. From this evidence, it is suggested that soil-forming periods during Late Siwalik time must have lasted on the order of ten thousand years, a factor not inconsistent with our determined fluvial cycle recurrence rate (see above).

The evidence that semiarid conditions prevailed in the Punjab 2.3 m.y. B.P. differs from some interpretations which hold that the absence of Upper Siwalik microflora is due to the onset of a colder, pre-glacial climate caused by uplift in the Himalayas and migration or death of pre-existing fauna (Nandi, 1975). However, Gill (1951) notes that there is no evidence of glacial deposits in the Upper Siwaliks. Further-more, the major weathering process in a frigid climate where organic material is decaying is that of decomposition, rather than the calcification observed here. A gley would develop, zones of leaching would be apparent, and the texture of the soil would not inherit that of the parent material (Hunt, 1972). Since none of these features were observed, glacial conditions are not supported by the paleosol evidence in the Upper Siwaliks. A similar conclusion was reached by Halstead and Nanda (1973), who studied sedimentologic and faunal evi-dence. They corroborate that the topography and general con-ditions must have resembled those currently existing on the Punjab.

Application to the
Siwalik Miocene Record

In spite of the rather abundant data available for the interpretation of part of the Upper Siwalik fluvial record,

the character of the streams responsible for Upper Siwalik
sedimentation remains problematic. Keller, et al. (1977)
concluded that the general alternation of fining-upward fluvial cycles in the Pabbi Hills was good evidence for Allen's
(1965b) fluvial system of high sinuosity streams, yet their
lateral extent was suggestive of relatively low sinuosity
streams. The fluvial stratigraphy of the Siwalik Group
observed elsewhere in the Potwar Plateau also may be at some
variance with Allen's models (Pilbeam, et al., 1979). The
presence of thick, well-developed vertical accretion units
implies the relatively prolonged existence of a differentiated flood plain environment, a feature of meandering
streams. However, the great lateral extent of lateral accretion bodies implies that streams were not confined to a
narrow meander belt but rather a broad alluvial piedmont much
like the present Indo-Gangetic Plain.

Fluvial cycles from the Himalayan molasse (Upper Siwalik
Sub-Group) in northern Pakistan provide the following
evidence:

1) In rapidly aggrading basins (i.e., sedimentation
   rates from 25 to 50 cm/1000 years in the present
   context) lithostratigraphic units can be shown
   to be approximately isochronous.

2) Differential subsidence in one traverse is approximately 2.5 m/km.

3) Differential subsidence is accompanied by an average
   rate of lateral accretion of nearly 0.2 km/1000
   years.

4) Repeated stream encounters at a depositional site
   may result in an easily differentiated series of
   single-storied fluvial cycles which, when considered in the context of time defined by paleomagnetic stratigraphy, may be the record of sedimentation events having a recurrence rate of some
   40,000 years.

5) Pedogenic alteration of floodbasin soils also
   appears to be developed over time periods in
   excess of ten thousand years.

6) Pedogenesis in Late Pliocene fossil soils (circa
   2.3 m.y. B.P.) is suggestive of semiarid alfisol
   development. These paleosols would develop under
   a mixed short grass/shrub and bush vegetative
   structure and respond to warm, but seasonally arid

climates which would average less than 80 to 100 cm of precipitation per year.

The approach demonstrated in the above discussion on Upper Siwalik alluvial environments has had limited testing on the earlier Chinji, Nagri and Dhok Pathan formations of the Central Potwar Plateau of Pakistan. The paleomagnetic geochronology of these sites is discussed in Barndt, et al. (1978), Johnson, N.M., et al. (1977) and Johnson, N.M., et al. (1980). The following discussion is based on these chronologies.

Paleosols were identified and sampled at seven stratigraphic levels between the upper Chinji and the top of the Dhok Pathan formation (Fig. 2, Table 1). Certain pedogenic features (e.g., color horizonation, color mottling, clay segregations, crotovina, fabric homogenization, etc.) were examined to determine the relative degree of pedogenesis attained in the various profiles prior to burial. These fabric properties, characteristic of the overbank mudstone facies, were considered in the context of their position within their respective fluvial cycle. Accordingly, these characteristics were used to identify the relative position of each paleosol profile on the ancient flood plain with which it was associated (Table 2).

Profile Descriptions

The following descriptions characterize several paleosol profiles sampled and analyzed:

CH-2-1

0-5 cm
: Greenish red claystone grading downward into siltstone. Crotovina and root casts. Finely laminated.

50-140 cm
: Dusky red (5 R 3/4) claystone. Mottled with green claystone.

140-190 cm
: Light olive-gray silty claystone. Mottled with green. Kankars present. Fining-upward.

190-250 cm
: Grayish red (10 R 4/2) claystone. Gradational contacts on top and bottom.

250-400 cm
: Light olive-gray claystone. Mottled with dark red claystone. Root casts, crotovina and kankars present. Possible $CaCO_3$ zone of enrichment at 340 cm.

Table 1. Lithostratigraphy, Location and Depositional Environment of Sampled Profiles.

Level	Stratigraphic Position	Location of Profile	Sedimentary Environment
KK-Gt-1A	Upper Dhok Pathan	Kundvall Kas, south of village of Dhok Pathan.	Distal floodbasin. Good paleosol development.
MK-12-1A	Middle Dhok Pathan	Malhuwala Kas across from village of Malhuwala.	Adjacent to river channel. Undisturbed laminations present.
MK-9-2	Lower Dhok Pathan	Malhuwala Kas, north of village of Malhuwala.	Middle of floodbasin, fairly proximal but with fair paleosol development.
MK-6-3A	Upper Nagri	100 m west of MK-6-1A on same stratigraphic level.	Distal floodbasin, possibly multi-storied paleosol.
MK-6-1A	Upper Nagri	Malhuwala Kas at Nagri-Dhok Pathan boundary.	Distal floodbasin, possibly multi-storied paleosol.
MK-3-1B	Middle Nagri	Malhuwala Kas - at same stratigraphic level as site #182.	Middles of flood basin, fairly proximal, with splays.

Table 1, continued.

Level	Stratigraphic Position	Location of Profile	Sedimentary Environment
DM-1-1	Middle Nagri	North of village of Dhok Mila, along Kas running by Dhok Mila.	Proximal flood basin. One possible paleosol is partially buried.
NG-1-1	Lower Nagri	North of Soan River along road between Talagang and Dhulian.	Distal flood basin. paleosol development.
CH-2-1	Upper Chinji	North of Soan River along road between Talagang and Dhulian.	Proximal flood basin with multi-storied paleosol development.

Table 2. Stratigraphic and environmental setting of selected paleosols from the Late Miocene Siwalik Group, Central Potwar Plateau.

Formation	River	Proximal Floodbasin	Mid-Floodbasin	Distal Floodbasin
Dhok Pathan		MK-12-1A	MK-9-1A	KK-G-1A*
				MK-6-1A*
8.2 m.y.#		MK-3-1B*		
Nagri		DM-1-1		
				NG-1-1*
10.1 m.y.#				
Chinji		CH-2-1		

*Na-dithionite-citrate extractable oxide distribution figured in report.
#Chronology from Johnson, et al. (1980).

The abundance of fabric features (crotovina, root casts, etc.) in the top 30 cm and the presence of undisturbed ripples below this level indicate that this soil was not exposed subaerially for a long period of time and consequently extensive fabric maturation did not occur. This suggests a stream-proximal sedimentary environment.

NG-1-1

0-450 cm   Dark reddish brown (10 R 3/6) claystone. Fining-upward from siltstone at 450 cm. Unindurated at base but becomes progressively indurated moving up in profile. Abundant root casts, crotovina and kankars. Root casts and crotovina filled with gray sand. Topped by thin siltstone bed, probably representing a distal-splay deposit.

450-600 cm   Sandstone, moderate yellowish brown (5 YR 5/4). Moderately indurated.

Extensive fabric maturation is indicated by the abundance of root casts, crotovina and kankars. The soil is stream-distal.

DM-1-1

0-100 cm   Grayish tan sandstone, possibly a splay deposit.

100-145 cm   Dark reddish brown (10 R 3/4) claystone. Possible pisoliths. Mottled coloration.

145-235 cm   Pale yellowish brown (10 YR 7/2) siltstone, becoming reddish-brown near bottom. Kankars may be present.

235-355 cm   Grayish red (10 R 4/2) claystone. Devoid of paleosol features. No bedding.

355-575 cm   Dark reddish brown (10 R 3/6) silty claystone. Well-indurated. Fining-upward from silty horizon near bottom.

575-700 cm   Dark reddish brown (10 R 3/6) claystone. Root casts and crotovina filled with gray sand. Kankars present.

A lack of fabric maturation features suggests that this soil occurred in a stream-proximal position.

MK-3-1B

3.5 m — Gray sandstone, poorly indurated.

0-230 cm — Pale brown (5 YR 5/2) claystone. Poorly indurated. Kankars.

230-250 cm — Pale yellowish brown (10 YR 6/2) sandstone Well-indurated. Top and bottom contacts gradational.

250-530 cm — Moderate yellowish brown (10 YR 5/4) claystone grading down into moderate brown (5 YR 4/4) claystone. Kankars, concentrated in zones. Fairly well-indurated.

530-600 cm — Yellowish brown sandstone, fining-upward into dark reddish brown claystone.

The presence of kankars indicates some fabric maturation but probably in a proximal environment.

MK-6-1A

2.5 m — Sandstone (subgreywacke). Poorly indurated.

0-600 cm — Gray reddish brown (10 R 4/4) claystone grading down into pale reddish brown (10 R 6/4) siltstone (at around 400 cm). Claystone has waxy texture near the top, and is indurated. Sandstone inclusions (indicating possible mudcracks), root casts and crotovina present.

600-1050 cm — Claystone (red grading down into tan) interbedded with tan-gray sandstone lenses.

Mud cracks, root casts and crotovina indicate substantial fabric maturation, probably in a distal environment.

MK-6-3A

0-225 cm — Grayish red (10 R 4/2) claystone. Waxy texture, indurated. Grades downward (75 cm) into moderate reddish brown (10 R 4/6) claystone. Texture not as waxy in lower part of profile. Sand ($CaCO_3$) fillings, root casts and crotovina present throughout.

Fabric maturation is relatively high indicating a relatively distal depositional site.

MK-9-2

0-225 cm   Moderate reddish brown (10 R 4/6) claystone Well-indurated. Kankars, root casts and crotovina present. No micro-structure or texture observed, massive. Mottled coloration between 150 and 225 cm.

225-350 cm   Moderate reddish brown (10 R 4/6) claystone Well-indurated. Abundant root casts and animal tracks. Wavy bedding-size of lamination increases with depth. Possible ancient surface at 250 cm.

350-500 cm   Moderate reddish brown (10 R 4/6) claystone. Large mottled spots occur around 350 cm. Possibly are ancient surface accumulations. Claystone becomes less indurated with depth.

Well-developed evidence of fabric maturation suggests intermediate-to-distal sedimentary environment.

MK-12-1A

Sandstone -   Massive, variable thickness. Light grayish brown (5 Yr 7/4).

0-60 cm   Pale reddish brown (10 R 5/4) claystone, well-indurated. Parallel laminations (less than .5 cm thick), undisturbed. No pedogenic features.

60-120 cm   Reddish tan sandstone. Poorly indurated.

120-350 cm   Pale red (5 R 6/2) claystone (slightly coarser than 0-60 cm section). Well-indurated. Slightly disturbed parallel laminations. No pedogenic. features.

350-650 cm   Light brown (5 YR 6/6) siltstone fining up to claystone. No pedogenic features.

Undisturbed ripples indicate that the profile was deposited in a proximal environment and that pedogenic modification is minor.

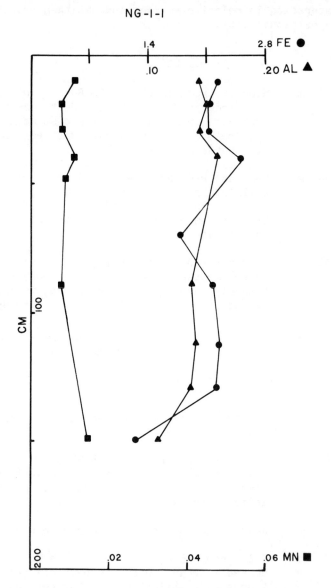

Figure 15.   Na-dithionite-citrate-extractable oxide distri-
bution for paleosol NG-1-1 sampled from 200 m
above the base of the Nagri Formation.
Site occurs 0.5 km east of the village of Dhok
Pathan-Dhulian road, at a point 6.5 km north of
Dhok Pathan, central Potwar Plateau.

KK-G-1A

0.7 m   Light brown (5 YR 6/4) sandstone, poorly indurated.

0.8 m   Moderate orangish brown (10 R 5/6) siltstone. Poorly sorted. Grades upward into sandstone.

0-250 cm   Dark reddish brown (10 R 4/4) claystone. Sand inclusions at 0-40 cm indicating mudcracks. Abundant root casts and kankars. Possible pisoliths. Zone of $CaCO_3$ accumulation at 200 cm.

250-370 cm   Olive-gray (5 YR 3/2) claystone. Fissile, with a waxy surface. Incipient kankars. Root casts.

370-450 cm   Pale yellowish brown (10 YR 6/2) sandstone. Very poorly sorted. Fines upward. Poorly indurated. Top contact gradational.

Root casts, pisoliths and other pedogenic features suggest that this soil was deposited in a distal environment where extensive pedogenesis could occur between depositional events (i.e., flooding).

The distribution of mobile oxides within the above soil profiles show definite zones of depletion and illuviation in the paleosol profiles sampled. Iron and aluminum oxides (sesquioxides) are leached from organic rich A-horizons and are deposited in the B-horizon in modern soils. Unfortunately, the organic material of the A-horizon is rarely preserved (Johnson, 1977) so that paleosol studies must be based on data from the remnant B-horizon. Zone of iron and aluminum illuviation are apparent in the distal profiles sampled (NG-1-1, MK-6-1A, and KK-G-1A) (Figs. 15, 16 and 17). While the more proximal profiles (DM-1-1, MK-3-1B (Fig. 18) and MK-12-1A) show neither depletion nor enrichment of the sesquioxides. This may be expected since pedogenetic change and associated leaching of sesquioxides is more extensive on distal parts of the flood plain where flooding and the addition of new parent material are not as frequent as on more proximal regions.

The relatively low concentrations of di-MnO indicate that these profiles were well-drained. Johnson (1977) reports that di-MnO distribution may reflect the relative

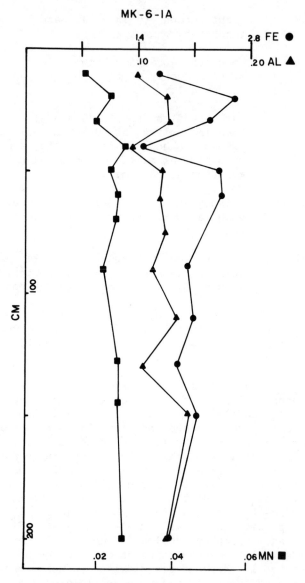

Figure 16.   Na-dithionite-citrate-extractable oxide distribution for paleosol MK-6-1A sampled from the
upper Nagri Formation,   60 m below the top
of the Formation. Site occurs along the right
bank of the Malhuwala Kas, 7.5 km downstream
from the village of Kamlial, eastern Khaur-
Dhulian anticline, central Potwar Plateau.

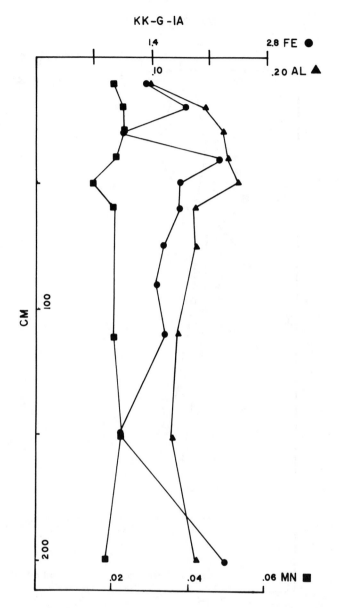

Figure 17.   Na-dithionite-citrate-extractable oxide distribution for paleosol KK-G-1A sampled from the
lower Dhok Pathan Formation, 100 m   above the
base   of   the   Formation.   Site occurs along
the right bank of the Kundvall Kas, 1.5 km upstream from its confluence with the Soan River,
2 km south of Khok Pathan village.

MK-3-1B

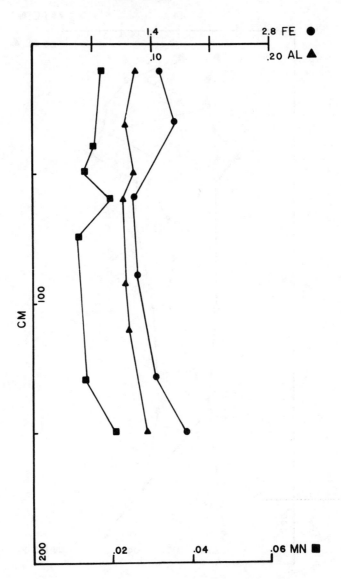

Figure 18.   Na-dithionite-citrate-extractable oxide distribution for paleosol MK-3-1B sampled from the
middle Nagri Formation, 120 m below the top of
the Formation.   Site   occurs along the right
bank of the Malhuwala Kas, 5.5 km downstream
from the village of Kamlial, eastern Khaur-
Dhulian anticline, central Potwar Plateau.

degree of leaching in various profile positions and that MnO accumulates in poorly drained parts of the profile.

Asamoa (1973) showed that free iron movement in soil profiles from Ghana is coupled to clay translocation in wetter regions and silt movement in drier regions. Blume and Schwertmann (1969) found that di-MnO moved in a soluble form independent of clay movement and that di-MnO concentration tended to peak at greater depths than di-Fe_2O_3 or di-Al_2O_3. This tendency is found to a limited degree in the distal profiles from Pakistan. In several cases, di-MnO maxima occur at the same depth as di-Fe_2O_3 minima (and vice versa) indicating that di-MnO accumulates in the parts of the profile where reducing conditions exist and di-Fe_2O_3 is most concentrated where conditions are strongly oxidizing.

Profiles NG-1-1 (Fig. 15) and MK-6-1A (Fig. 16) have several di-Fe_2O_3 maxima that may represent fossil B-horizons. This type of multi-storied profile development, also seen in the Pliocene sequence discussed above, is common in proximal flood plain soils where there is rapid addition of new parent material to the top of the profile. High concentrations of di-MnO provide evidence that these profiles were frequently flooded. Several of the other profiles including DM-1-1 and KK-G-1A also show possible multi-storied development. Mottling of red and green colors below 50 cm coupled with high di-MnO concentrations below 75 cm indicates a shallow water table that probably fluctuated in depth beneath the soil surface. This is commonly observed in proximal flood plain sites.

Discussion

The paleosols studied were formed on an alluvial flood plain during a period when the Potwar Plateau was experiencing the relatively high rate of aggradation of 50 cm/1000 years (Johnson, N.M., et al., 1980). Consequently these alluvial soils were not exposed subaerially for long periods of time before burial and only immature profiles were developed and preserved. Since pedogenesis is a function of many variables including climate, time and parent material (Blume and Schwertmann, 1969), caution should be placed on the interpretation of paleoclimates based on alluvial soil profile data (Birkeland, 1974). Distal profiles lend themselves best to paleoclimate interpretations since more time is recorded in pedogenesis compared to proximal areas where sedimentation rates are higher, burial takes place sooner, and parent material is more variable.

From the data collected it is apparent that sesquioxide movement is a direct function of the degree of pedogenesis (which primarily is a function of time) and that maximum translocation (i.e., movement of clay and sesquioxides from the A-horizon to B-horizon) occurs in profiles that have undergone intense pedogenesis. In all cases, maximum oxide translocation is found in distal profiles which were exposed subaerially for the longest period of time. The intensity of pedogenesis in these alluvial paleosols in a function of time allowed for pedogenesis (see also Johnson, 1977).

The three profiles (NG-1-1, MK-6-1A and KK-G-1A) developed on distal parts of flood plains were analyzed for paleoclimatic interpretations. The Fe_2O_3 distribution patterns of profile NG-1-1 (Fig. 15) and also of DM-1-1 are similar to the distribution of Fe_2O_3 in an Ultisol from Costa Rica described by Martini and Macias (1974). Characteristics of both profiles include a relatively constant concentration of Fe_2O_3 between the surface and a depth of about 30 cm. This is followed by a gradual increase in the Fe_2O_3 concentration, which eventually peaks and gradually drops. The soil sampled by Martini and Macias was developed beneath an artificial savanna which replaced a previously tropical rain forest. The mean annual precipitation is 2490 mm and there is one dry season each year.

The mobile oxide distribution patterns for profile MK-6-1A (Fig. 16) and KK-G-1A (Fig. 17) are similar to data presented by Asamoa (1973) for soils developed under a semi-deciduous forest in Shana. The features of these soils include an Fe_2O_3 minimum in the A-horizon followed by a sharp maxima in the upper part of the B-horizon (between 10 cm and 30 cm). The Fe_2O_3 maxima in the forest soils tend to be greater than those found in comparable savanna soils. Data presented by Juo, et al. (1974) and by Ahmad and Jones (1969) also suggest that mobile oxide distribution patterns of profiles MK-6-1A and KK-G-1A indicate that these soils developed under a semi-deciduous forest which received between 1270 mm and 1775 mm of precipitation per year. Asamoa found that clay translocation followed Fe_2O_3 movements, indicating that the two phenomena may be linked to each other.

A comparison of profile MK-6-1A and KK-G-1A (which seem to have experienced the same intensity of pedogenesis) shows that the average and maximum Fe_2O_3 content of KK-G-1A is less than that of MD-G-1A and the depth of maximum concentration is less in KK-G-1A than in MK-6-1A. This trend represents a more arid climate during the formation of KK-G-1A since the depth of leaching decreases as the climate becomes drier and the Fe_2O_3 concentration decreases.

Late Miocene Siwalik Geochronology

Several magnetic polarity sections have been established in the Potwar Plateau and include the Chinji, Nagri and Dhok Pathan stratotypes. Preliminary data on the geochronologies which are established were reported by Barndt, et al. (1978). Recently these data have been greatly refined and supplemented such that a reliable magnetic stratigraphy can be erected for most of the Siwalik Group (Johnson, N.M., et al., 1980).

The dominant feature of this chronology is a long normal polarity zone which is always associated with the Nagri Formation. This persistent and conspicuous normal polarity zone contains one volcanic ash bed in the Nagri stratotype which we have dated at 9.5 $^+$ .6 m.y. B.P. The normal polarity zone thus can be identified as magnetic Chron 9. Additionally, the radiometrically dated Upper Siwalik ashes discussed above constrain the magnetics determined from certain of these stratigraphic sections. The magnetic polarity stratigraphy of three of these sections, including the Kotal Kund site in the Eastern Salt Range (see Fig. 14a, site KK4, the locality from which the example of Upper Siwalik pedogenesis was selected) and the Mulhawala Kas section in the Khaur-Dhulian anticline, south of the village of Kamlial, in the central Potwar Plateau, has been correlated with the accepted magnetic polarity time scale (Johnson, N.M., et al., 1980). As a result, the ages of the stratigraphic units in the central Potwar can be defined as follows: The Chinji, Nagri and Dhok Pathan formations have nominal age ranges of 13.1 to 10.1, 10.1 to 8.2 and 8.2 to 5.0 m.y., respectively. The Dhok Pathan interval in its type area south of the village of Dhok Pathan in the central Potwar records only a portion of post Nagri time: along the Kundvall Kas section and related sites to the east, the youngest strata are not younger than circa 7.4 m.y. (Johnson, N.M., et al., 1980). The area to the southeast in the vicinity of Kotal Kund (Fig. 14a) contains a more complete record.

Late Miocene Environmental Change

As a result of the chronologies established for the Chinji, Nagri and Dhok Pathan formations, the paleosols studied in the central Potwar may be put into a proper chronometric sequence (Table 2).

Data from Pakistani paleosols suggest that during Chinji

and early Nagri time (circa 10.5 to 9.0 m.y. B.P.) the
Siwalik landscape was developed under a tropical forest which
was being replaced by a semi-deciduous forest and savanna.
The climate was warm year-round and rainfall averaged more
than 1300 mm per year, with definite wet and dry seasons.
Soils were typical ferruginous soils with and without
plinthite (ironstone) concretions. These incipient latosols
were undoubtedly only one component of a mosaic of soil types
likely to be found in the Siwalik alluvial plain. Shifting
sedimentation patterns above allowed for the development of
many stream-proximal to stream-distal soil types which would
record differently the overall climatic controls on the
landscape (Table 2). These factors are discussed elsewhere
(Johnson, G.D., 1977; Visser and Johnson, 1978).

Starting in early Dhok Pathan time (circa 8.0 m.y.
B.P.), the shale-to-sandstone ratio increases rapidly indicating lower gradient stream systems and the establishment of
a well-developed meander-belt facies. The climate apparently
became more seasonally arid during the Dhok Pathan, as is
indicated by fabric and sesquioxide distributions, and the
average temperature remained high. Average rainfall probably
dropped to 1250 mm by the end of the Dhok Pathan and the
vegetation changed from a woody semi-deciduous forest to a
more sparse open savanna or grassland. Gallery forests may
have still occurred along the river at this time.

A contemporaneous Siwalik site at Haritalyangar, H.P.
(India) has been similarly studied for evidence of fossil
soil-forming processes (Johnson, 1977). The paleosols of
that sequence are roughly from the same stratigraphic interval as paleosol MK-12-1A and reflect continued development of
ferruginous oxisol or incipient laterite development in the
Siwalik alluvial landscape well into the upper portion of
magnetic Chron 7. The evidence from these soil types and the
chronology we have established indicate a rather persistent
macro-climate (evidenced from the paleosol record) for this
portion of the latest Miocene (Johnson, G.D., et al., in
preparation).

The chronology of these soils is as follows:

Haritalyangar paleosol #60:   circa 7.0 m.y. B.P.

Haritalyangar paleosol #54:   circa 7.5 m.y. B.P.

        MK-12-1A   :   circa 7.7 m.y. B.P.

        MK-9-1A   :   circa 8.3 m.y. B.P.

KK-G-1A   :   circa   8.6 m.y. B.P.

MK6-1A   :   circa   8.8 m.y. B.P.

MK3-1B   :   circa   9.0 m.y. B.P.

DM-1-1   :   circa   9.8 m.y. B.P.

NG-1-1   :   circa 10.0 m.y. B.P.

CH-2-1   :   circa 10.9 m.y. B.P.

A more fully developed chronology for the Haritalyangar/ Potwar correlation is discussed elsewhere (Johnson, G. D., in preparation).

Nandi (1975) has stated that palynological data suggest a change from a warm, humid environment in early Siwalik time (Chinji Formation) of the Punjab in India to a cooler, drier climate by late Siwalik time. The middle Siwalik Group (Nagri and Dhok Pathan formations) represents a time of change. Paleosols from Pakistan also suggest that the climate was becoming more seasonally arid during the middle Siwaliks, particularly after magnetic Chron 8 (circa 7.9 m.y. B.P.). However, it should be reiterated that the data from the Pakistani and Indian paleosols do not indicate a large drop in average temperature as Nandi suggests. The climatic change during this time period may be related to a pulse in the Himalayan orogeny that could have changed climatic patterns (Nandi, 1975). Mammalian fossil assemblages analyzed by Tattersall (1969a and b) and Prasad (1971) also support a more arid climate by the end of Dhok Pathan deposition. Moonen, et al. (1978) suggested that the Dhok Pathan fauna may illustrate evidence of a changing biotope, as evidenced by the establishment of open habitat fauna quite different from the more closed habitat fauna of the earlier Chinji and Nagri. Pilbeam, et al. (1979) analyzed the faunal stability in herbivores from the Chinji, Nagri and Dhok Pathan in the central Potwar and recognized little change occurring in the community structure during magnetic Chrons 11, 10 and 9 (approximately the Chinji and Nagri formations in the central Potwar). The herbivore composition began changing in the medial part of magnetic Chron 8 (lower Dhok Pathan Formation) suggesting more open habitat. In all cases, however, a mixed woodland-grassland habitat is indicated, with certain faunistic components representative of riparian fringe (closed habitat) communities and others of more stream-distal (open habitat) communities. This change in fauna in Dhok Pathan time (circa 7.9 m.y. B.P.) is consistent with our analysis above, but must represent only the initial evidence

of change to more arid conditions. The implications of continued aridity, or more realistically a greater seasonality
to rainfall distribution, is that the transition from the
more ferruginous to the more calcareous paleosols must be
evident.

Within rocks of the Dhok Pathan Formation, in the vicinity of Kotal Kund where a most complete record is preserved
(Johnson, N.M., et al., 1980), evidence for the initiation of
aridisol development is found in rocks of lower magnetic
Chron 5 and upper magnetic Chron 6. This is approximately
6.0 m.y. B.P. In rocks of the upper most Dhok Pathan (above
the Bhandhar Bone Beds of Pilgrim, 1913) and in the remainder
of the Siwalik sequence, abundant evidence of kankars and
other disseminated calcareous concretions are found in the
overbank mudstones of these fluvial sediments. In Pakistan,
this trend certainly continued into latest Pliocene time as
evidenced by the paleosol profile development in the Kotal
Kund region and elsewhere in the Eastern Salt Range. Thus
the cumulative evidence from the Upper Siwalik paleosols is
that this drying trend has continued, resulting in more
seasonally arid conditions and a more open vegetative physiognomy in late Siwalik time.

Faunal Implications

The first appearance, thus far established, for hominoid
primates in the Siwaliks of the Potwar Plateau region is 10.0
to 10.5 m.y. B.P. (discussed in Pilbeam, et al., 1977b, 1979,
and Johnson, N.M., et al., 1980). In particular these
ramapithecine hominoids occur throughout a variety of sites
in the Potwar and the Eastern Salt Range in Pakistan and
throughout the vicinity of Haritalyangar, Himachal Pradesh
and the Jammu foothill belt in India.

This first appearance may subsequently be documented in
older rocks but at present represents our best estimate based
on modern stratigraphically documented specimens and certain
specimens from historic collections which have precise locality data associated with them.

The upper stratigraphic limit of hominoid primate
occurrences in the Siwaliks, thus far determined, is within
middle Dhok Pathan rocks: the stratigraphic occurrence of
Gigantopithecus bilaspurensis from Haritalyangar being the
stratigraphically youngest primate thus far recovered. The
historic and recent collections of Siwalik ramapithecines and
dryopithecines all occur in strata which are older than this
(Pilbeam, et al., 1977b, 1979; Johnson, N.M., et al., 1980).
The exact chronology of these hominoid local ranges is

discussed elsewhere (Johnson, N.M., et al., 1980) but in all cases the last appearance datum is older than 6.0 m.y. B.P. for these sites in Pakistan and northwestern India (Johnson, G. D., et al., in preparation).

Habitat Exclusion

The last appearance, although admittedly based on scant fossil material, seems to coincide with the change from tropical ferruginous soils which predominate in the earlier strata to semiarid and arid calcareous soils which characterize the younger strata. This coincidence may provide evidence of exclusion of the hominoid habitat during late Siwalik time in Pakistan and northwestern India about six million years ago.

The ecologic gradient of the extant vegetative community along the sub-Himalayan foothill belt from east to west in northern India today is very steep. Proceeding from a lush, tropical forest physiognomy in Arunchal Pradesh in the east through a tall grass/dense forest structure in western Nepal and in the vicinity of Dehra Dun in Uttar Pradesh, the general vegetative structure gives way quite rapidly to the grass/thorn bush savanna of the Punjab. Further west still, a true arid zone can be seen in the Western Salt Range and Trans-Indus Ranges.

Clearly this ecologic gradient or vegetative catena with associated soils quite possibly existed in the past. The arid paleoenvironments of late Dhok Pathan and late Siwalik time may have restricted or excluded the habitat of hominoid primates. If this is true, by analogy to extant systems, this exclusion should be most severe in the geographic northwest or west of the cline. An exception to this of course is the probable orographic placement of favorable habitats in adjacent upland areas to the north of the Siwalik molasse basin. Sites within that portion of the Siwalik sedimentary basin, in reality the Himalayan foredeep or molasse basin, which are presently accessible lie at an undetermined distance to the south of the evolving Himalayan mountain front (and its potentially equitable habitats).

One could speculate that sites recording the favorable hominoid habitat from latest Miocene, Pliocene and Pleistocene aged strata should lie to the east of the area discussed in this text in a more equitable portion of the environmental cline. As such, sites from the contemporary Nepali Siwaliks and further east may record the equitable hominoid habitat which was lost in the Potwar about six million years ago.

Acknowledgements

Appreciation is extended to the personnel of Peshawar University and the Geological Survey of Pakistan for their continued interest in and support of this research. J. Barndt, N.M. Johnson, H.M. Keller, M. Khan, D.R. Pilbeam, R.G.H. Raynolds and P.K. Zeitler all contributed variously to this project. This research was supported by U. S. National Science Foundation grants GF-44168, GA-43445, EAR 76-20717, EAR 78-03639, Smithsonian Institution grant F-R4-601117 (to D.R. Pilbeam), Peshawar University, and Dartmouth College.

This work has been conducted under a collaborative project among Peshawar University, Dartmouth College, Lamont-Doherty Geological Observatory, the Geological Survey of Pakistan and the University of Arizona.

References

Ahmad, N., and Jones, R.L., 1969a, A plinthaquilt of the Aripo Savannas, North Trinidad: I. Properties of the soil and chemical composition of the natural vegetation: Soil Sci. Amer. Proc., v. 3, p. 761-765.

Ahmad, M., and Jones, R.L., 1969b, A plinthaquilt of the Aripo Savannas, North Trinidad: II. Mineralogy and genesis: Soil Sci. Soc. Amer. Proc., v. 33, p. 765-768.

Ahmad, M., Ryan, J., and Paeth, R.C., 1977, Soil development as a function of time in the Punjab river plains of Pakistan: Soil Sci. Soc. Amer. Proc. (Madison), v. 41, p. 1162-1166.

Allen, J.R.L., 1965a, Fining-upwards cycles in alluvial successions: Geol. Jour., v. 4, p. 229-246.

Allen, J.R.L., 1965b, Review of recent alluvial sediments: Sedimentology, v. 5, p. 89-191.

Allen, J.R.L., 1970, Studies in fluviatile sedimentation: a comparison of fining-upward cyclothemes, with special reference to coarse member composition and interpretation. Jour. Sediment. Petrol., v. 40, p. 298-323.

Allen, J.R.L., 1974, Studies in fluvial sedimentation: Implications of pedogenic carbonate units, Lower Old Red Sandstone, Anglo-Welsh Outcrop: Geol. Jour. (Liverpool), v. 9, p. 181-204.

Asamoa, G.K., 1973, Particle size and free iron oxide distribution in some latasols and groundwater laterites of Ghana: Geoderma., v. 10, p. 285-297.

Barndt, J., 1977, The magnetic polarity stratigraphy of the type locality of the Dhok Pathan faunal stage, Potwar Plateau, Pakistan (M.A. thesis): Dartmouth College, Hanover, New Hampshire, U.S.A.

Barndt, J., et al., 1978, The magnetic polarity stratigraphy and age of the Siwalik Group near Dhok Pathan Village, Potwar Plateau, Pakistan: Earth and Planet. Letters, v. 41, p. 355-364.

Birkeland, P.W., 1974, Pedology, Weathering and Geomorphology Research, Oxford University Press, New York, 295 p.

Blume, H.P., and Schwertmann, V., 1969, Genetic evaluation of profile distribution of aluminum, iron and manganese oxides: Soil Sci. Soc. Amer. Proc., v. 33, p. 438-444.

Bridge, J.S., and Leeder, M.R., 1979, A simulation model of alluvial stratigraphy: Sedimentology, v. 26, p. 617-644.

Buol, S.W., Hole, F.D., and McCracken, R.J., 1973, Soil Genesis and Classification, Iowa State University Press, Ames, 360 p.

Buursink, J., 1971, Soils of Central Sudan, Grafisch Bedryf Schotamus and Jens Utrecht N.V., Utrecht, p. 249.

Cotter, G. deP., 1933, The geology of the part of the Attock District west of longitude 72°45' E: Memoirs of the Geological Survey of India, v. 55, p. 63-156.

deTerra, H., and Teilhard de Chardin, P., 1936, Observations on the Upper Siwalik Formation and later Pleistocene deposits in India: Am. Phil. Soc. Proc., v. 76, p. 791-822.

Fermor, L.L., 1931, General report for 1930: Geol. Surv. India Records, v. 65, p. 118-125.

Gee, E.R., 1945, The age of the Saline Series of the Punjab and of Kohat: Proc. Nat. Acad. Sci. India, v. 14, p. 269-312.

Gee, E.R., 1947, Further note on the age of the Saline Series of the Punjab and of Kohat: Proc. Nat. Acad. Sci. India, v. 16, p. 95-116.

Gerasinov, I.P., 1971, Nature and originality of paleosols, in Yaalon, D.H., Paleopedology - origin, nature and dating of paleosols: Intl. Soc. of Soil Sci. and Israel Universities Press, Jerusalem, p. 3-13.

Gile, L.H., 1970, Soils of the Rio Grande Valley border in southern New Mexico: Soil Sci. Soc. Amer. Proc. (Madison), v. 34, p. 465-472.

Gile, L.H., and Hawley, J.W., 1966, Periodic sedimentation and soil formation on an alluvial fan piedmont in southern New Mexico: Soil Sci. Soc. Amer. Proc. (Madison), v. 30, p. 261-268.

Gill, W.D., 1951, The stratigraphy of the Siwalik series in the Northern Potwar, Punjab, Pakistan: Quat. Jour. (London), v. 107, p. 375-394.

Gingerich, P.D., 1969, Markov analysis of cyclic alluvial sediments: Jour. Sediment. Petrology, v. 39, p. 330-332.

Halstead, L.B. and Nanda, A.C., 1974, Environment of deposition of the Pinjor Formation, Upper Siwaliks, near Chandigarh: Indian Geol. Assoc. Bull., v. 6, n. 1, p. 63-70.

Holmgren, G.G.S., 1967, A rapid citrate-dithionite extractable iron procedure: Soil. Sci. Soc. Amer. Proc., v. 31, p. 210-211.

Hunt, C.B., 1972, Geology of Soils: Their Evolution, Classification, and Uses, W.H. Freeman and Company, San Francisco, 344 p.

Jenny, H., 1941, Calcium in the soils: III. Pedologic relations: Soil Sci. Soc. Amer. Proc. (Madison), v. 6, p. 27-35.

Johnson, G.D., 1977, Paleopedology of Ramapithecus-bearing sediments, North India: Geol. Rund., v. 66, p. 192-216.

Johnson, G.D., and Vondra, C.F., 1972, Siwalik sediments in a portion of the Punjab re-entrant: the sequence at Haritalyangar, District Bilaspur, H.P.: Himalayan Geology, v. 2, p. 118-144.

Johnson, G.D., et al., 1979, Magnetic reversal stratigraphy and sedimentary tectonic history of the Upper Siwalik Group Eastern Salt Range and Southwestern Kashmir, in Farah, A. and DeJong, K.A. eds., Geodynamics of Pakistan: Geol. Surv. Pakistan, Quetta, p. 149-165.

Johnson, N.M., Opdyke, N.D., and Lindsay, E.H., 1975, Magnetic polarity stratigraphy of Pliocene-Pleistocene terrestrial deposits and vertebrate faunas, San Pedro Valley, Arizona: Geol. Soc. Amer. Bull., v. 86, p. 5-12.

Johnson, N.M., et al., 1977, Magnetic polarity stratigraphy of the Middle Siwalik Group, Potwar Plateau, Pakistan: Geol. Soc. Amer. Abstr. with Prog., v. 9, p. 1039-1040.

Johnson, N.M., et al., 1980, Magnetic polarity stratigraphy and ages of the Chinji, Nagri and Dhok Pathan Formations of the Potwar Plateau, Pakistan. in preparation.

Juo, A.S.R., Moormann, F.R., and Maduakor, H.O., 1974, Forms and pedogenetic distribution of extractable iron and aluminum in selected soils of Nigeria: Geoderma., v. 11, p. 167-179.

Kamei, T., 1977, The Siwalik Series and the Plio-Pleistocene boundary: The Quaternary Res. (Daiyonki-Kenkyu), v. 15, p. 181-185.

Keller, H.M., 1975, The magnetic polarity stratigraphy of an Upper Siwalik sequence in the Pabbi Hills of Pakistan (M.A. thesis): Kresge Library, Dartmouth College, Hanover, New Hampshire, U.S.A.

Keller, H.M., et al., 1977, Magnetic polarity stratigraphy of the Upper Siwalik deposits, Pabbi Hills, Pakistan: Earth Planet. Sci. Lett., v. 36, p. 187-201.

Krynine, P.D., 1937, Petrography and genesis of the Siwalik Series: Amer. Jour. of Sci., v. 34, p. 422-466.

Leakey, L.S.B., 1969, Ecology of North Indian Ramapithecus: Nature, v. 223, p. 1075-1076.

Leeder, M.R., 1975, Pedogenic carbonates and flood sediment accretion rates: a quantitative model for alluvial arid-zone lithofacies: Geol. Mag., v. 112, p. 257-270.

Leeder, M.R., 1978, A quantitative stratigraphic model for alluvium, with special reference to channel deposit density and interconnectedness: Canadian Soc. Petrol. Geol. Mem., v. 5, p. 587-596.

McKeague, J.A., Brydon, J.E., and Miles, N.M., 1971, Differentiation of forms of extractable iron and aluminum in soils: Soil Sci. Soc. Amer. Proc. (Madison), v. 35, p. 33-38.

Martini, J.A., and Macias, M., 1974, A study of six latosols from Costa Rica to elucidate the problems of classification, productivity and management of tropical soils: Soil Sci. Soc. Amer. Proc. (Madison), v. 38, p. 644-652.

Miall, A.D., 1974, Markov chain analysis applied to ancient alluvial plain succession: Sedimentology, v. 20, p. 347-364. Soil Sci. Soc. Amer. Proc., v. 38, p. 644-651.

Miall, A.D., 1977, A review of the braided-river depositional environment: Earth Sci. Rev., v. 13, p. 1-62.

Moody-Stuart, M., 1966, High and low sinuosity stream deposits, with examples from the Devonian of Spitsbergen: Jour. Sediment. Petrol., v. 36, p. 1102-1117.

Moonen, J.J.M., et al., 1978, A comparison of larger fossil mammals in the stratotypes of the Chinji, Nagri and Dhok Pathan Formations (Punjab, Pakistan): Kon. Neder. Akad. Weten. Proc., v. 81, p. 425-436.

Nandi, B., 1975, Palynostratigraphy of the Siwalik Group of the Punjab: Himalayan Geology, v. 5, p. 411-427.

Opdyke, N.D., 1972, Paleomagnetism of deep-sea cores: Rev. Geophys. Space Phys., v. 10, p. 213-249.

Opdyke, N.D., et al., 1977a, Paleomagnetism of the Upper Siwalik sediments of Pakistan: Trans. Amer. Geophys. Union, v. 58, p. 379.

Opdyke, N.D., et al., 1977b, The magnetic polarity stratigraphy of the Upper Siwalik sediments of northeast Pakistan and the age of the transition between the Pinjore and Tatrot land mammal faunas: Geol. Soc. Amer. Abst. with Prog., v. 9, p. 1120-1121.

Opdyke, N.D., et al., 1979, Magnetic polarity stratigraphy and vertebrate paleontology of the Upper Siwalik Subgroup of Northern Pakistan: Palaeogeog. Palaeoclim. and Palaeoecol., v. 27, p. 1-34.

Pascoe, E., 1930, General report for 1929: Geol. Surv. India Records, v. 63, p. 125-141.

Papadakis, J., 1969, Soils of the World, Amsterdam, Elsevier, 208 p.

Perkin Elmer Corp., 1968, Analytical methods for atomic absorption spectrophotometry, Perkin Elmer Corp.

Pilbeam, D.R., et al., 1977a, Geology and palaeontology of Neogene strata of Pakistan: Nature, v. 270, p. 684-689.

Pilbeam, D.R., et al., 1977b, New hominoid primates from the Siwaliks of Pakistan and their bearing on hominoid evolution: Nature, v. 270, p. 689-695.

Pilbeam, D.R., et al., 1979, Miocene sediments and faunas of Pakistan: Postilla, no. 179, p. 1-45.

Pilgrim, G.E., 1913, The correlation of the Siwaliks with mammal horizons of Europe: Geol. Surv. India Records, v. 43, p. 264-325.

Pons, L.J., and Zonneveld, I.S., 1965, Soil ripening and soil classification: Int. Inst. Land Reclam. and Improv. Publ., v. 13, p. 1-128.

Prasad, K.N., 1971, Ecology of the fossil hominoidea from the Siwaliks of India: Nature, v. 232, p. 413-414.

Rieken, F.F., and Poetsch, E., 1960, Genesis and classification considerations of some prairie-formed soil profiles from local alluvium in Adair County, Iowa: Iowa Academy of Sci., v. 67, p. 268-276.

Ruellan, A.,1971, The history of soils: some problems of definition and interpretation, in Yaalon, D.H., ed., Paleopedology - origin, nature and dating of paleosols: Intl. Soc. of Soil Sci. and Israel Univ. Press, Jerusalem, p. 3-13.

Sehgal, J.L., and Stoops, G., 1972, Pedogenic calcite accumulation in arid and semiarid regions of the Indo-Gengetic alluvial plain of Erstwhile Punjab (India) - their morphology and origin: Geoderma (Amsterdam), v. 8, p. 59-72.

Sehgal, J.L., Sys, C., and Bhumbla, D.R., 1968, A climatic soil sequence from the Thar Desert to the Himalayan Mts. in Punjab, India: Pedologie (Ghent), XVIII, v. 3, p. 351-373.

Selley, R.C., 1970, Studies of sequence in sediments using a simple mathematical device: Geol. Soc. Quart. Jour., v. 125, p. 557-581.

Sidhu, P.S., and Gilkes, R.J., 1977, Mineralogy of soils developed in alluvium in the Indo-Gangetic Plain (India):

Soil Sci. Soc. Amer. Proc. (Madison), v. 41, p. 1194-1201.

Smithson, F., 1956, Plant opal in soil: Nature (London), v. 178, p. 107.

Soil Survey Staff, 1951, Soil Survey Manual: USDA Handbook No. 18, U.S. Govt. Printing Office, Washington, D.C., 503 p.

Soil Survey Staff, 1975, Soil Taxonomy: A Basic System of Soil Classification for Making and Interpreting Soil Surveys: Agriculture Handbook No. 436, Soil Conservation Service, U.S.D.A.

Steila, D., 1976, The Geography of Soils, Prentice-Hall, Englewood Cliffs, N.J., p. 222.

Tattersall, I., 1969a, Ecology of North India Ramapithecus: Nature, v. 221, p. 451-452.

Tattersall, I., 1969b, More on the ecology of North Indian Ramapithecus: Nature, v. 224, p. 821-822.

Teruggi, M.E., and Andreis, R.R., 1971, Micromorphological recognition of paleosolic features in sediment and sedimentary rocks, in Yaalon, D.H., ed., Paleopedology - origin, nature and dating of paleosols: Intl. Soc. of Soil Sci. and Israel Universities Press, Jerusalem, p. 161-172.

Visser, C.F., and Johnson, G.D., 1978, Tectonic control of Late Pliocene molasse sedimentation in a portion of the Jhelum re-entrant, Pakistan: Geol. Rund., v. 67, p. 15-37.

Wadia, D.N., 1945, A note on the repeated overthrusts of the Cambrian rocks on the Eocene in the northeastern part of the Salt Range: Proc. Nat. Acad. Sci. India, v. 14, p. 214-221.

Carl F. Vondra, Mark E. Mathisen,
Daniel R. Burggraf, Jr., Erik P. Kvale

8. Plio-Pleistocene Geology of Northern Luzon, Philippines

Abstract

The Cagayan Valley basin is a north-south trending interarc basin in which 10,000 meters of marine, transitional marine, and fluvial sediments have been deposited since the Oligocene. The Plio-Pleistocene sediments have been divided into two lithostratigraphic units, the Ilagan Formation and the Awidon Mesa Formation, which include five major lithofacites. The facies and respective depositional environments are: 1) the interbedded fine-grained sandstone and mudstone facies--delta front distal bar and distributary mouth bar, 2) the lenticular trough cross-bedded medium-grained sandstone and siltstone facies--delta plain distributary channel, levee, and flood basin, 3) the polymictic conglomerate, trough cross-bedded sandstone and claystone facies--low energy fluvial channel and floodplain, 4) the polymictic clast supported conglomerate and sandstone facies--high energy channel bar and gravel sheet, and 5) the massive matrix-supported pebble-to-boulder conglomerate and pyroclastic sandstone facies--ignimbrites and lahars. The facies record a regression of the Pliocene sea as the basin filled with detritus from the surrounding volcanic arcs. A delta complex prograded to the north as sediment was received from a large north flowing meandering stream system. Braided streams transported coarser detritus from the volcanic arcs to the valley and formed an alluvial fan complex at the base of the active Cordillera Central volcanic arc. In the Pleistocene, the Cagayan Valley alluvial plain supported a tree savannah ecosystem with large grazing mammals which had migrated from Asia via a land-bridge. Paleolithic stone tools found in surface associations with fossilized remains of the Pleistocene fauna suggest that hominids also migrated to the Philippines in the Pleistocene. Thus far, however, no hominid skeletal evidence has been found in the sediments.

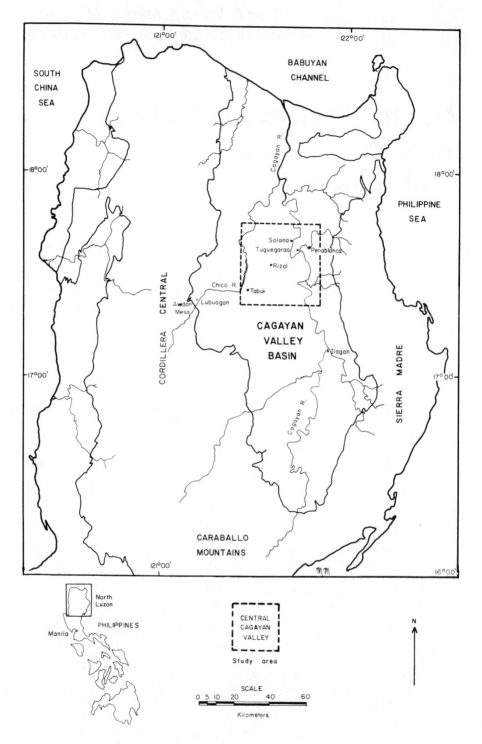

Figure 1. Index map of central Cagayan Valley.

Introduction

The Plio-Pleistocene terrestrial sediments of the central Cagayan Valley, northern Luzon, the Philippines, have been the focus of numerous archaeological and paleontological studies. Preliminary investigations by von Koenigswald (1958) and the Philippine National Museum (Fox, 1971; Lopez, 1972; Fox and Peralta, 1974) have documented the occurrence of Paleolithic pebble-cobble tools of the Asian chopper-chopping tool tradition, flakes, and fossilized remains of a Pleistocene vertebrate fauna. The artifacts and fossils have been found at sixty-eight sites on the eroded tops and slopes of grassy, often gravel-covered anticlinal hills. Most of the artifacts are surface finds of unknown age that have not, as yet, been directly associated with the vertebrate fossils which are thought to be Middle Pleistocene in age (Fox and Peralta, 1974).

In China (Black, 1933; Movius, 1949) and Indonesia (Deterra 1943; von Koenigswald and Ghosh, 1973) the same chopper-chopping tool tradition has been dated by direct association with an extinct Middle Pleistocene fauna and Homo erectus. Previous workers in the Cagayan Valley (Fox and Peralta, 1971) have postulated that there may also be a direct association between the Philippine artifacts and extinct Middle Pleistocene fauna indicating that Homo erectus inhabited the Philippines during the Middle Pleistocene.

To better understand the significance of the Paleolithic artifacts and Pleistocene fauna with regard to Philippine and Southeast Asian prehistory, a cooperative project between archaeologists and geologists of the Philippine National Museum, archaeologists of the University of Iowa and geologists of Iowa State University was organized and initiated in 1978. Field studies utilizing an interdisciplinary approach are being conducted to define the Plio-Pleistocene terrestrial sequence in the Cagayan Valley basin, demonstrate the in situ association of artifacts and Pleistocene fauna, and document the age of the artifacts, specific animal species, present, and the Plio-Pleistocene environments of the valley.

This report describes the Plio-Pleistocene geology and paleoenvironments of the central Cagayan Valley. The central Cagayan Valley, between 121°25' and 121°50' east longitude and 17°20' and 17°45' north latitude is an area of over 2,000 square km which encompasses the towns of Penablanca, Tuguegarao, Solana, Rizal, and Tabuk (Fig. 1). The data and interpretations included in this report are based primarily on field observations. Detailed laboratory analyses

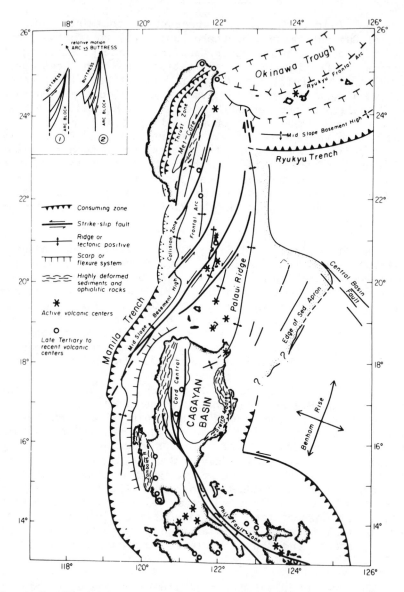

Figure 2.   Tectonic map of the Luzon Taiwan region. The inset schematically shows a possible mechanical solution to explain the internal deformation within the arc system (after Karig, 1973).

of the collected samples are being initiated to supplement the fieldwork.

Geologic Setting

The Cagayan Valley is a major north-south trending intermontane basin approximately 250 km long and 80 km wide which developed in the mobile belt bordering the Asian mainland. The tectonic elements of this belt in the Luzon-Taiwan region are shown in Figure 2. The Manila trench occurs to the west of Luzon while the Philippine trench occurs to the east and south. A faulted submarine ridge runs between Luzon and Taiwan to the north. The Cagayan basin itself is bordered to the north by the sea and flanked by mountain ranges to the east, west, and south (Figure 1). These are the Sierra Madre, Cordillera Central, and Caraballo mountains which are composed of basaltic and andesitic rocks of volcanic arc origin (Durkee and Pederson, 1961; Santos-Ynigo, 1966; Hamilton, 1973; Karig, 1973; Murphy, 1973).

The Cagayan basin may be classified as a back arc or interarc basin (Fig. 3) based on interpretations published by Karig (1973), Murphy (1973), and Bowin et al. (1978). The Paleogene Sierra Madre formed as an island arc in response to convergence along a west-dipping subduction zone east of Luzon. During the Late Oligocene, the polarity of the arc was reversed as subduction to the east ceased and a new east-dipping subduction zone formed to the west of Luzon. The Cagayan basin began to form following this reversal with the uplift of the Cordillera Central volcanic arc during the Late Oligocene-Early Miocene. Thus, the basin formed as a back arc basin behind the Neogene Cordillera Central which remained active through the Pleistocene, or an interarc basin between the active Cordillera Central and the inactive Sierra Madre volcanic arc. Roedder (1977) has presented a different interpretation of the evolution of northern Luzon. He considers the area to be the result of an arc-arc collision.

Sedimentation began in the Oligocene with the deposition of over 8,000 m of Oligocene and Miocene turbidites and submarine debris flows on volcanic and metasedimentary basement rocks of the subsiding Cagayan basin (Durkee and Pederson, 1961; Tamesis, 1976). With the uplift and formation of the Cordillera Central, the basin gradually became asymmetrical, and the basin axis migrated from west to east (Fig. 4) during the Miocene and through the Pleistocene (Christian, 1964; Caagusan, 1978). During the Plio-Pleistocene, 400 to 2,000 m of transitional marine and fluvial sediments of the Ilagan and Awidon Mesa formations were deposited (Corby et al., 1951; Durkee and Pederson,

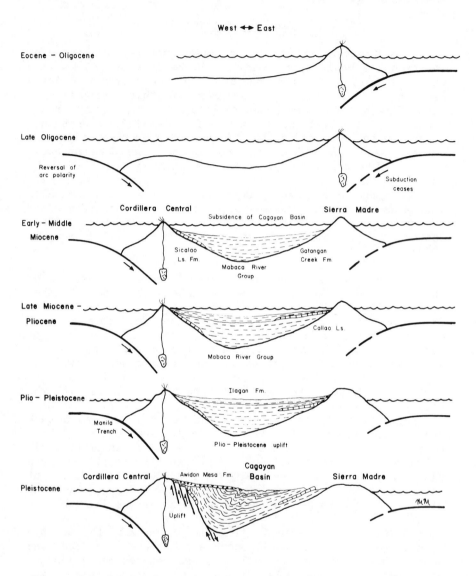

Figure 3. Schematic island arc evolution of the Cagayan
basin.

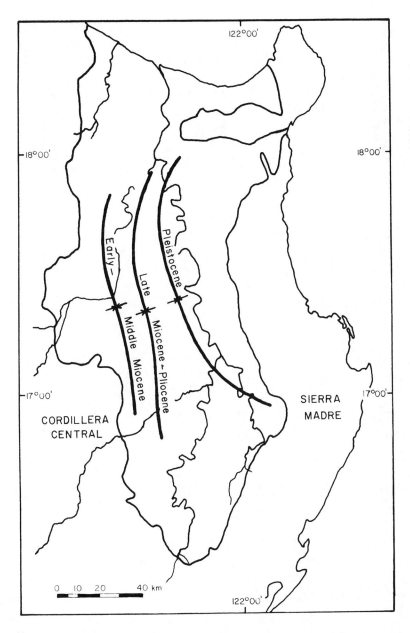

Figure 4. Migration of Cagayan basin axis in Miocene through Pleistocene Epochs (after Caagusan, 1978). Reprinted by permission.

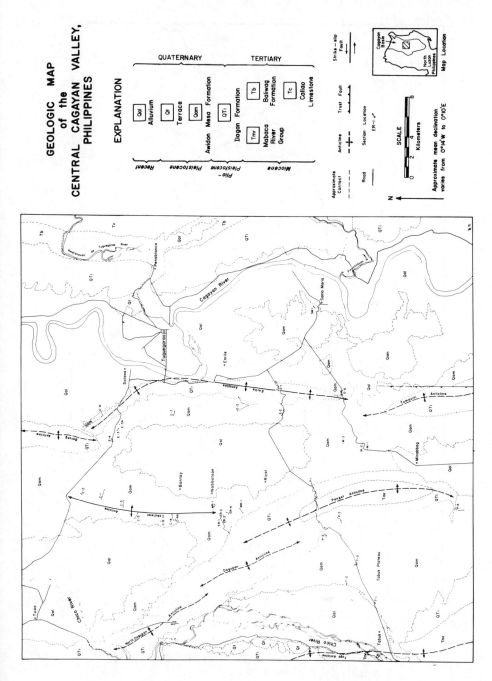

Figure 5. Geologic map of the central Cagayan Valley.

1961; Tamesis, 1976). The Awidon Mesa Formation, a 300 m thick sequence of Pleistocene pyroclastic and fluvial sediments, conformably overlies the Ilagan Formation in the valley but unconformably overlies folded Miocene and Pliocene strata in the foothills of the Cordillera Central.

The Cagayan basin sediments were folded in the Pleistocene forming the Cagayan Valley anticlinal belt which trends north-south along the western side of the valley. The folds, which do not persist with depth, were attributed to gravity sliding as the Cordillera was uplifted (Christian, 1964). Consideration of the active tectonic setting and the conditions necessary for gravity sliding (Kehle, 1970) suggest that gravity sliding is the most likely cause of folding in this area. The folding and subsequent erosion is recorded by an intraformational unconformity in the Pleistocene Awidon Mesa Formation.

The dates assigned to the Plio-Pleistocene fluvial sediments and fauna at present are based on a minimal amount of evidence. It is this terrestrial sequence which is of major importance to this study.

Stratigraphy

Approximately 10,000 m of sediment have been deposited in the Cagayan basin since the Oligocene. The predominantly clastic succession was deposited on volcanic and metasedimentary rocks of the pre-Oligocene basement complex (Durkee and Pederson, 1961; Tamesis, 1976). The Oligocene and Miocene sediments, which were only briefly examined during reconnaissance, are of marine origin and are overlain by transitional marine and fluvial Plio-Pleistocene deposits.

Thirty-five stratigraphic sections were measured, described, and sampled throughout the study area to provide a basis for interpreting the Plio-Pleistocene stratigraphy and paleoenvironments of the central Cagayan Valley. Fourteen of the sections were measured in the Cabalwan Anticline area where most of the National Museum archaeologic and paleontologic sites occur. A generalized geologic map of the central Cagayan Valley (Fig. 5) has been prepared utilizing 1:50,000 topographic maps as base maps. Since the use of aerial photographs was restricted by the government, some data are inferred from previous maps published by Durkee and Pederson (1961) and Christian (1964).

As a result of a renewed oil exploration effort, the stratigraphic nomenclature for the basin has been recently revised (Tamesis, 1976). The revised nomenclature for the Oligocene and Miocene rocks contrasts significantly with that

Table 1. Stratigraphic nomenclature of the Cagayan Valley.

Age			Corby et al., 1951	Kleinpell, 1954	Durkee and Pederson, 1961	Tamesis, 1976	Environments
Quaternary	Recent	Pleis-tocene	Ilagan Formation / Late Tertiary Volcanics / Magapit Ls.	Ilagan Formation	Awidon Mesa Formation	Awidon Mesa Formation	Fluvial
Tertiary	Pliocene		Tuguegarao Sandstone		Ilagan Formation	Ilagan Formation	Deltaic
	Miocene	Upper	Callao Ls.	Callao Ls.	Baliwag Formation	Cabagan Formation	Prodelta
		Middle	Lubuagan Coal Measures	Lubuagan Coal Measures	Mabaca River Group / Callao Ls.	Callao Ls.	Shelf to Deep Bathyal
		Lower	Ibulao Ls.	Ibulao Ls.	Gatangan Creek Formation / Sicalao Ls.	Lubuagan Formation / Dumata Formation	
	Oligo-cene						Supralittoral to Littoral
	Pre-Oligocene		Basement – Basic igneous volcanic rocks and metasediments				

proposed by Corby et al. (1951), Kleinpell (1954), and Durkee
and Pederson (1961) as indicated in Table 1. The Pliocene
Ilagan Formation and Pleistocene Awidon Mesa Formation have
largely been ignored by petroleum geologists due to their
lack of oil potential. Both the Ilagan and Awidon Mesa formations were examined during this study because the contact
relationship between the Pliocene and Pleistocene rocks had
not been determined in the basin. The nomenclature proposed
by Durkee and Pederson (1961) is being used at this time
because the nomenclature mentioned by Tamesis (1976) has not
been adequately defined for other workers to use. The
Plio-Pleistocene stratigraphy will be described in detail
following a brief review of the Miocene stratigraphy. The
generalized stratigraphy of the Plio-Pleistocene terrestrial
deposits, which are the subject of archaeological and paleontological studies, is illustrated in Figure 6.

Miocene, Sicalao Limestone

This formation is named after the Sicalao River on the
western side of the valley (Durkee and Pederson, 1961). Five
hundred forty-six meters of thin to massively bedded
fossiliferous carbonates were measured at the type section
along Anaguan Creek, 3 km north-northwest of the Rizal,
Cagayan.

The Sicalao Limestone can be traced nearly continuously
along the western margin of the valley where it overlies
mafic igneous rocks of the basement complex. In some areas,
however, the Sicalao Limestone is absent due to post-Sicalao
faulting and erosion. The formation was assigned an Early
Miocene age based on its orbitoid fauna (Durkee and Pederson,
1961).

Miocene, Mabaca River Group

The Mabaca River Group, which is typically exposed along
the Mabaca River, Kalinga-Apayao, includes all strata on the
western side of the valley which overlie the Sicalao
Limestone or basement and underlie the Pliocene Ilagan
Formation (Durkee and Pederson, 1961). The group consists of
three formations, the Asiga Formation, the Balbalan
Sandstone, and Buluan Formation, which together form a thick
sequence of lutites, interbedded arenites, and some
pyroclastic deposits. Based on our reconnaissance
observations, these deposits may be interpreted as shelf to
deep bathyl clays and turbidites. These strata were
previously referred to as the Lubuagan Formation by Corby et
al. (1951) who did not designate a type area or section. The
Lubuagan area is a structurally complicated region where it
would be difficult to measure a complete stratigraphic
section. The most complete section of the Mabaca River Group

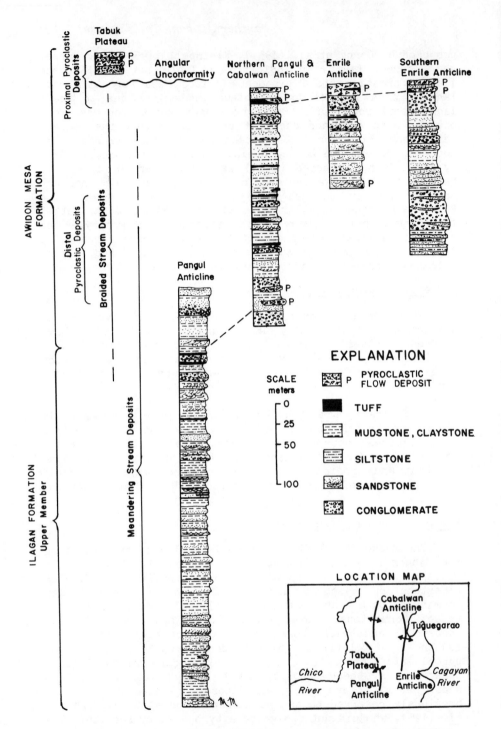

Figure 6. Generalized Plio-Pleistocene stratigraphy of the central Cagayan Valley terrestrial deposits.

is 8,200 m thick and was measured 6 km east of Lubuagan between Toloctoc and Naneng. Based on the microfauna, the Mabaca River Group is considered to be of Early to Late Miocene or Miocene-Pliocene age (Durkee and Pederson, 1961). The Mabaca River Group is exposed along the western margin of the study area at the Chico River bridge and also at Pangul anticline which is breached to the Buluan Formation.

Miocene, Gatangan Creek Formation

On the east side of the Cagayan Valley, andesite flows of the basement complex are overlain by graywackes and claystones of the Gatangan Creek Formation (Durkee and Pederson, 1961). The formation is named after Gatangan Creek, due east of Cabagan, Isabela. Here the formation is 1,010 m thick and overlain by the Callao Limestone. The graywackes and claystones of the Gatangan Creek are here interpreted to be shelf and deep bathyl clay and turbidite deposits similar to those of the Mabaca River Group. Foraminiferal assemblages indicate that the formation is Early to Middle Miocene in age, correlative with the lower part of the Mabaca River Group (Durkee and Pederson, 1961).

Miocene, Callao Limestone

The Callao Limestone overlies the Gatangan Creek Formation on the east side of the valley. Corby et al. (1951) named the Callao Limestone and designated Barrio Callao, Cagayan as the type area. At Callao Canyon along the Pinacanaun de Tuguegarao River, the formation consists of 540 m of reef carbonates (Durkee and Pederson, 1961). The Callao was considered Middle Miocene in age by Durkee and Pederson (1961) based on foraminifera collected at the type locality. The Philippine National Oil Company, on the basis of more complete paleontological evidence, has redated the Callao as Late Miocene and Pliocene in age (E.V. Tamesis, PNOC, Manilla, personal communication, 1978).

Miocene, Baliwag Formation

Overlying the Callao Formation and underlying the Ilagan Formation is the Baliwag Formation, a valley-forming claystone which was named and described by Vergara et al. (1959). It varies from 100 m thick along the eastern margin of the study area to 420 m thick at the type area along the Baliwag River near the Cagayan-Isabela provincial boundary. The Baliwag was considered to be Late Miocene by Durkee and Pederson (1961) and correlative with the Buluan Claystones of the Mabaca River Group. Based on its stratigraphic occurrence, lithology, primary structures, and faunal content, the Baliwag is interpreted to be a prodelta deposit.

Pliocene, Ilagan Formation

The name Ilagan was proposed by Corby et al. (1951) for a sequence of poorly cemented sandstones 200 to 400 m thick along the valley floor. No type area or type section was designated. It has generally been agreed by other workers that the name refers to outcrops of Plio-Pleistocene clastic sediments along the Ilagan River, south of Ilagan, Isabela Province, which exhibit the typical fluviatile depositional nature of the formation (Kleinpell, 1954; Durkee and Pederson, 1961; Tamesis, 1976). No detailed description of the formation has been previously given because significant lateral lithological variations occur over short distances and there was no economic reason to study it.

The Ilagan Formation was studied in detail in the central Cagayan Valley to gain a better understanding of basin evolution and the contact relationship with the overlying Awidon Mesa Formation. In this area the Ilagan Formation, as mapped by Durkee and Pederson (1961), consists of a lower 150-310 m thick sequence of thin interbedded sandstones, siltstones and mudstones, and an upper 570 m thick sequence of thicker, more massive, cross-bedded conglomerates, sandstones, siltstones, and claystones. Based on a regional reconnaissance of the valley, Kleinpell (1954) suggested the formation be divided into two members, a lower mudstone member, here called the Lower Member, and an upper sandstone, the Upper Member.

Lower Member: The Lower Member is composed of thin, interbedded sandstones, siltstones, and mudstones which conformably overlie Late Miocene and Pliocene claystones of the Baliwag Formation on the east side of the valley and the Mabaca River Group on the west side. The basal contact is gradational and mostly covered but distinct at the base of escarpments formed by the resistant Ilagan sandstones. The upper contact is the base of the first thick tabular body of trough cross-bedded sandstone which fines upward to siltstone and a thick claystone. The dark greenish gray (5 GY 4/1) to pale yellowish orange (10 YR 8/6), fine- to very finely grained sandstones and siltstones are laterally continuous, well-sorted, and usually exhibit parallel laminations, lenticular bedding or flaser bedding or small-scale trough cross-bedding. In the upper part of the Lower Member, more massive lenticular, medium-to-coarse-grained sandstones up to 9 m thick occur. The bedding is often indistinct, but small and large-scale trough cross-bed sets up to 50 cm thick are common. Planar cross-beds up to 30 cm thick were also present at one locality. Blackish red (5 R 2/2) calcareous and ferruginous disk- to blade-shaped concretions up to boulder size are present along numerous horizons within the Lower

Member. Gastropods and pelecypods are locally abundant in
sandstones while plant fragments and burrows are common
throughout the member. Numerous shark teeth found on the
surface appear to have been eroded from the Lower Member.

Upper Member: Plio-Pleistocene conglomerates, sandstones, siltstones, and claystones that occur between the
first tabular sandstone body which fines upward to claystone
and the first pyroclastic flow deposit or quartz granule-
bearing conglomerate form the Upper Member of the Ilagan
Formation. The Upper Member outcrops along the flanks of
Enrile and Pangul anticlines in the central Cagayan Valley.
The entire Upper Member is exposed at Pangul Anticline where
it attains a thickness of 570 m. Only the upper part of the
Member is exposed at Enrile Anticline. The conglomerates,
sandstones, and claystones form laterally extensive sheet-
like deposits which are commonly traceable for at least
several kilometers.

The sediments of the Upper Member typically occur in
fining-upward sequences. In the lower part of the member,
poorly sorted, medium to coarse-grained, pale yellowish
orange (10 YR 8/6) sandstones grade upward to well-sorted,
fine-grained sandstones and siltstones which are overlain by
massive, pale olive (10 Y 6/2) claystones which may sometimes
be sandy. There is also a coarsening-upward throughout the
member as polymictic granule to cobble conglomerates become
more common and thicker in the upper part of the member. The
polymictic conglomerates, which are dominantly clast
supported, are composed primarily of porphyritic andesite,
basalt, metasedimentary clasts, and chert. Primary structures characteristically grade upward from large-scale trough
cross-bed sets of up to 1 m thick in the conglomerates and
coarse sandstones to small-scale trough cross-beds in the
finer sandstones. Heavy mineral-rich layers and grain size
variations accent the cross-bedding. Climbing ripple lamination may also occur in the finer sandstones. Scour-and-fill
is common in the conglomerates and sandstones while load
structures, convolute bedding, and calcareous concretions
occur occasionally. The siltstones are usually thin and
characterized by parallel lamination. The overlying claystones are massive, commonly reaching 20 m in thickness. The
sandstones usually are thick, reaching 12 m, while the
conglomerates only thicken in the upper portion of the
member. The fining-upward sequences often attain a thickness
of 20 m.

Thin, lenticular, very light gray (N8) to very pale
orange (10 YR 8/2) tuffs occur throughout the Upper Member
associated with the upper sandstones, siltstones, and
claystones of the fining-upward sequences. Along the north-

west flank of Pangul Anticline, the tuffs occur in three distinct tuffaceous intervals which commonly contain per-mineralized logs and abundant leaf impressions, primarily ferns. Trace fossils such as burrows and root casts occur in the tuffs as well as in the associated siltstones and claystones. No megafossils have been found in the Upper Member of the Ilagan.

Pleistocene, Awidon Mesa Formation

The Awidon Mesa Formation was named and described by Durkee and Pederson (1961). They used the term to describe Middle Pleistocene tuffaceous sediments of a dacitic type which are characterized by the presence of bipyramidal quartz (generally less than 5%), euhedra of hornblende, and sodic feldspar. The formation unconformably overlies folded strata of Miocene age at its type locality, Awidon Mesa, in the Cordillera Central near Lubuagan, Kalinga-Apayao Province. It attains a thickness of 300 m at Awidon Mesa and discon-tinuously extends out into the Cagayan Valley, where it overlies older tuffaceous sediments of the Ilagan Formation (Durkee and Pederson, 1961). The contact relationship be-tween the Awidon Mesa and the Ilagan Formation in the valley has not been previously studied. One of the major objectives of the fieldwork to date has been to determine the character and distribution of the Awidon Mesa Formation in the Cagayan Valley.

The formation grades from a thick, massive, valley-filling sequence of dacitic pyroclastic flow deposits in the mountains at the type area to a sequence of thinner pyroclastic flow deposits and tuffs that are interbedded with tuffaceous fluvial deposits along the western side of the Cagayan Valley. The term pyroclastic flow deposit is used as defined by Lajoie (1979) to refer to deposits of volcanic clasts that were ejected from vents and transported en masse on land or in water. The basal contact of the Awidon Mesa grades laterally from an unconformable relationship with the underlying folded Miocene sediments in the mountains to a conformable relationship with the Upper Member of the Ilagan in the Cagayan Valley. The base of the oldest dacitic pyroclastic flow deposit or the first quartz granule-bearing conglomerate is here designated as the contact between the Awidon Mesa Formation and the underlying Ilagan in the Cagayan Valley.

The textures of the pyroclastic deposits vary con-siderably from very well sorted, white (N9) tuffs to breccias with a poorly sorted, light gray (N7) dacitic sand matrix and angular to rounded clasts up to boulder size. In some of the younger deposits in the Valley, boulders several meters in diameter occur, the largest having a long axis of 5.5 meters. The clasts in the coarser pyroclastic deposits are dominantly matrix-supported. Orientation varies as clasts in some deposits are oriented parallel to the bedding while clasts in other deposits have random orientations. The clasts weather in relief giving each deposit a characteristically knobby, grayish red (5 R 4/2) surface.

Compositionally, the matrix of the coarse-grained pyroclastic deposits is uniform throughout the area studied, but the dominant lithology of the clasts varies. All the deposits have a light gray (N7) dacitic sand matrix with a low clay content. The moderately indurated concrete-like appearance of the pyroclastic deposits has probably resulted from devitrification of the clay-sized ash as noted by Schminke (1967) in lahars from Washington State. The matrix also contains a minor but significant amount of bipyramidal quartz as observed by Durkee and Pederson (1961). The quartz commonly forms a sparkling erosional residue on the surface where the Awidon Mesa Formation is present. The mineralogy of the pyroclastic deposits is presently being examined in detail to determine if significant variations occur which would be useful for distinguishing between deposits. Clast lithology varies with the age and location of the pyroclastic deposits. The youngest deposits which form the Tabuk Plateau at the base of the Cordillera Central contain primarily subangular, equant, very light gray (N8) dacite and light gray (N7) andesite clasts. Some light greenish gray (5 GY 8/1) and pale red (5 R 6/2) andesite-dacite clasts also occur with minor amounts of basalt. The older pyroclastic deposits exposed to the east along the flanks of Cabalwan, Pangul, and Enrile anticlines contain primarily rounded equant- to disk-shaped pumice clasts. Minor amounts of basalt fragments are also sometimes present.

A variety of sedimentary structures occurs in the pyroclastic deposits. The basal contact of the pyroclastic deposits is usually sharp with little or no relief. However, scour-and-fill structures may be present at the base of some deposits. Many of the coarser deposits are massive or may exhibit graded bedding or reverse graded bedding. In many outcrops several flow units may be present separated by sharp contacts. Small- and large-scale trough cross-bedding occurs at the base or top of many of these units. Massive beds or beds with normal or reverse graded bedding commonly grade to cross-bedded deposits laterally indicating reworking by streams. The massive-appearing tuffs are also usually reworked by streams as indicated by the presence of faint climbing ripple laminations.

Outcrop characteristics of the pyroclastic rocks were

carefully examined to provide a basis for the local correlation of strata and possible subdivision of the Awidon Mesa Formation for documentation of archaeological˙ sites or fossil localities. The tuffs are usually lenticular and thin but may attain a thickness as great as 2 m. They are difficult to use for correlation as they usually pinch out over a short distance or are covered. The coarsest pyroclastic flow deposits are more valuable for correlation because thicknesses are nearly uniform. Two 1 to 3 m thick, coarse-grained pyroclastic flow deposits occur along the flanks of Cabalwan Anticline. Similar 1 to 3 m thick deposits occur along the eastern flank of Enrile Anticline and both flanks of Pangul Anticline. A thicker, 3 to 20 m massive pyroclastic sequence occurs along the west flank of Enrile Anticline. It usually forms a resistant cliff but is occasionally eroded or covered by alluvium. At this time, thickness, sheet-like geometry, stratigraphic position, and grain size are the major distinguishing characteristics of the pyroclastic rocks that are useful for correlating exposures along the flanks of the anticlines and in some cases across synclines. Color and primary structures have not yet proved useful for distinguishing between deposits. However, mineralogic studies in progress of the pyroclastic deposits may indicate that there are distinct differences that may be applied to problems of correlation.

The pyroclastic deposits of the Awidon Mesa Formation are interbedded in the Cagayan Valley with thicker deposits of polymictic conglomerates, sandstones, siltstones, mudstones, and claystones. Fining-upward sequences occur throughout the formation but are most common in the lower part. Granule to pebble conglomerates and large-scale trough cross-bedded sandstones fine upward to siltstones and claystones. The sandstones and conglomerates, up to several meters thick, are thinner and more discontinuous than the sandstones and conglomerates of the underlying Ilagan Formation. Lenticular sandstones commonly pinch out within a few hundred meters. Like the sandstones, the associated claystones of the Awidon Mesa Formation are not generally as thick as those of the Ilagan. The upper part of the Awidon Mesa Formation is characterized by thicker, massive polymictic cobble-boulder conglomerates, conglomeritic sandstones, and sandstones. Individual conglomerates are up to 22 m thick while conglomeritic sequences attain a thickness of 50 m. The conglomerates vary from matrix-supported to clast-supported and are often imbricated indicating flow from the west. The disc- to equant-shaped, rounded clasts are composed primarily of basalt, porphyritic andesite, metasedimentary clasts, chert, and a small amount of jasper. Siltstones and claystones occur in this upper conglomeritic sequence but are usually not more than several meters thick and are often truncated by lenticular conglomerates and sandstones.

The conglomerates of the Awidon Mesa Formation in the central Cagayan Valley increase in thickness and clast size from the north to the south. In the north, along the flanks of Cabalwan and Enrile anticline, the conglomerates are up to 6 m thick and composed of pebble- to cobble-size clasts. Individual conglomerates form resistant ridges along the flanks of the anticlines and are usually traceable for at least 1 km. Extensive limonite-coated lag gravels from erosion of these conglomerates cover many of the hills along the flanks of the anticlines. To the south, at the Wanawan Ranch along the northeastern flank of Pangul Anticline, it was not possible to measure the thickness of the conglomerates because of cover, but an increase in grain size was observed along with an abundance of equant, rounded, small boulders. Farther to the south, the thickest conglomerates occur at the southern nose of Enrile and Pangul anticlines. These conglomerates are up to 50 m thick and coarsen upward from cobbles and small boulders to very large boulders of dacite. The boulders also increase in size to the west from large boulders at Enrile anticline to very large boulders at the south nose of Pangul anticline, the largest of which has a long axis of 5.5 m.

The lag gravels in the northern part of the study area contain tektites, naturally occurring glasses of possible extraterrestrial or impact origin (Barnes and Barnes, 1973; O'Keefe, 1963). Ninety-one tektites were found during the 1979 field season while thousands had been previously recovered by National Museum field workers. The tektites were once abundant (over 500 were collected at one locality on the Wanawan Ranch by the Philippine National Museum, personal communication, Lito Soriano, 1979) but are now scarce as a result of collections made by local people and visitors. Most of the tektites are equant in shape and rounded indicating abrasion by stream transport. Teardrop and dumbbell shaped tektites do occur, however. The grooved and pitted surface sculpture (Figure 7) is typical of Philippine (Beyer, 1961) and Southeast Asian tektites (O'Keefe, 1963), and preliminary tektite chemistries determined at Iowa State University are correlative with published Philippine tektite compositions (Table 2) (Barnes and Barnes, 1973, p. 106; O'Keefe, 1963, p. 69). The stratigraphic occurrence of all tektites recovered during this study has been noted along with the location of all tektite localities. At this time it appears that all of the tektites have eroded out of a conglomerate or conglomeritic sequence along the flanks of Cabalwan, Enrile, and Pangul anticlines. Excavations are planned in an attempt to find tektites in situ to confirm

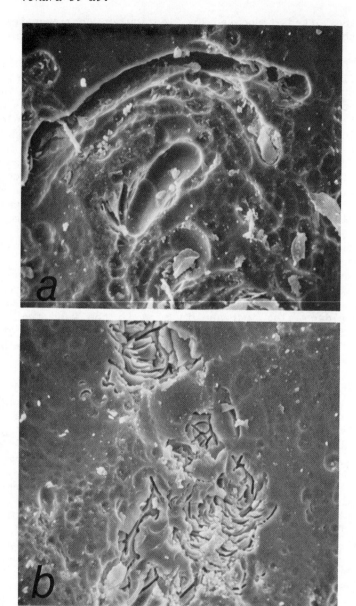

Figure 7. SEM photographs of Cagayan Valley tektite surface
          features
          a. 300x; note pitting along arcuate traces.
          b. 300x; note etching along concentric, semi-
              concentric, and linear traces

Table 2. Major and minor element composition of Cagayan Valley tektites and average Philippine tektite composition

	Tektite Composition[a]	Average Philippine Tektite Composition[a]	
	Cagayan Valley	Barnes & Barnes (1973, p. 106)	O'Keefe (1963, p. 69)
SiO_2	68.82	71.21	70.80
Al_2O_3	14.31	12.57	13.85
FeO_T	5.30	5.51	4.93
CaO	2.49	3.19	2.89
MgO	2.27	2.90	2.75
Na_2O	1.62	1.52	1.78
K_2O	2.50	1.93	2.35
MnO	0.10	0.11	0.09
TiO_2	0.84	0.89	0.75

[a]Oxide weight percent.

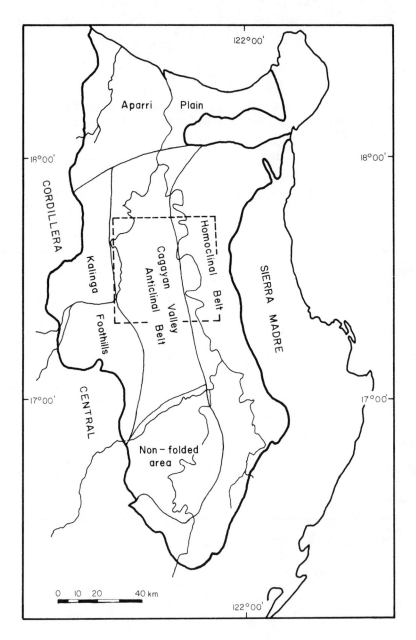

Figure 8.   Cagayan Valley structural provinces

their stratigraphic occurrence.   In addition, the tektites will be dated radiometrically.   Such dates must be used with caution since tektites may be easily reworked and deposited with younger sediments as noted by Harrisson (1975).

The Awidon Mesa Formation is very fossiliferous in contrast to the underlying Pliocene Ilagan Formation. Disarticulated remains of a variety of fossil vertebrates: elephant, rhinoceros, carabao, pig, deer, turtle, and crocodile were found on the surface and in situ in sandstones, conglomerates, and claystones.   Most of the vertebrate fossils were found along the northwest flank of Enrile Anticline, the northeast flank of Pangul Anticline, and along both flanks of Cabalwan Anticline. Besides vertebrates, permineralized wood and leaf impressions are common.   Trace fossils, such as burrows and root casts, are present in many of the siltstones, claystones, and sandstones, and molds are common in pyroclastic flow deposits.

Structure and Tectonic History of the Central Cagayan Valley

The central part of the Cagayan Valley may be divided into three major structural provinces (Fig. 8).   Along the western side of the valley is a very strongly folded belt (Caagusan, 1978) which has been referred to as the Kalinga foothills (Durkee and Pederson, 1961).   It is characterized by broad synclines and steeply folded anticlines which have been uplifted and eroded, exposing Miocene turbidites of the Mabaca River Group.   A homoclinal belt is present along the eastern margin of the valley.   Distinct homoclinal ridges, dipping from 10° to 14° to the west, are formed by the resistant Ilagan and Callao formations.   A moderately folded belt   (Caagusan,1978)   trends   north-south   throughout   the central part of the Cagayan Valley and is here referred to as the Cagayan Valley anticlinal belt as named by Durkee and Pederson (1961).   This belt contains about twenty major asymmetrical doubly plunging anticlines which are separated by broad, flat, alluviated synclinal valleys.   The Cabalwan, Pangul, and Enrile anticlines occur in the central part of the belt where most of the fieldwork for this study was concentrated.   Extensive exposures of the Ilagan and Awidon Mesa formations occur along the flanks of these anticlines which vary from about 15 to 25 km in length.   The Pangul and Enrile anticlines, are overturned on the east flank similar to several other anticlines in the belt (Durkee and Pederson, 1961; Christian, 1964).   The Cabalwan Anticline is also asymmetrical but has a steeper dip of 38° along the western flank and more gentle dips of 7° to 10° along the east flank.

The tectonic history of the Cagayan Valley has been interpreted by Christian (1964).   Following Miocene marine

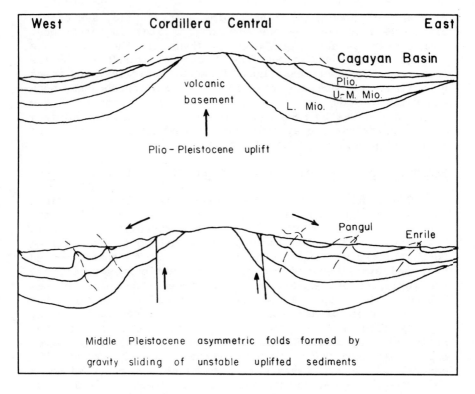

Figure 9. The formation of folds by Middle Pleistocene gravity sliding (after Christian, 1964)

sedimentation, regional uplift occurred while the Pliocene and Early Pleistocene sediments were being deposited. Very little tilting or compressional deformation accompanied this phenomenon. During the Pleistocene, the eastern margin of the basin rose enough to form the homoclinal belt and expose the Late Miocene Callao Limestone. Christian (1964) suggests that Middle Pleistocene oversteepening of the Cordillera Central led to mass gravity failure of the sediments along the mountain front and the formation of strongly asymmetrical to overturned anticlines in the Cagayan anticlinal belt (Fig. 9). This gravity-sliding mechanism was proposed because, according to seismic and drilling data, the folds do not persist with depth and compressive features such as thrust faults and folds are not known in the Sierra Madre or Cordillera Central foothills.

A review of the mechanics of gravity sliding supports the suggestion by Christian (1964) that this is the most likely cause of folding in the Cagayan Valley. The formation of folds by gravity sliding is dependent on many factors. Viscosity variations within a sequence of sediments and the thickness of the beds are two of the more important factors which control the development of a decollement zone and the velocity of the slide (Kehle, 1970). In the Cagayan Valley the shales of the Mabaca River Group form a 8,000 m thick zone where decollement could occur. Kehle (1970) notes that the contribution of gravity increases with 1) the dip of the potential slide mass because the component of gravity in the direction of motion increases with dip and 2) the depth of burial of the potential decollement zone as the shear stress in a dipping bed increases linearly with depth of burial. As the Cordillera Central was uplifted, the dip of the Plio-Pleistocene terrestrial sediments began to steepen over an already thick and deeply buried potential decollement zone. The slope at which gravity sliding may begin depends on the fluid pressure (Hubbert and Rubey, 1959; Lemoine, 1973) and the weight or thickness of the slide mass (Goguel, 1948). Various calculations have been worked out indicating that a 2- to 4-km-thick layer will flow on slopes ranging from 2° to 5° (Goguel, 1948; Hubbert and Rubey, 1959). Hose and Danes (1973) have noted that once sliding is initiated, it may create a mass deficiency in the hinterland and a mass surplus along the leading edge. The net effect in the rear would be uplift in the autochthon to provide a slight local gradient increase. Other factors which may contribute to a potential gravity slide are changing temperatures which may cause variations in pore pressure (Lemoine, 1973) and earthquakes (Pierce, 1973) which may help to trigger a slide. Along the Cordillera Central, higher temperatures and earthquakes would have been significant because of the volcanic and tectonic activity associated with the active island arc.

Altogether, conditions in the Cagayan Valley were favorable to initiate gravity sliding of the Plio-Pleistocene fluvial sediments during the Pleistocene, and asymmetric to over-turned folds occurring in the valley today are the fold type to be expected as a result of gravity sliding.

Stratigraphic fieldwork has provided evidence by which a firm date for the Pleistocene folding can be fixed.   An intraformational unconformity is present along the western flank of Pangul Anticline. Tuffaceous sediments of the lower part of the Awidon Mesa Formation are folded as part of the Anticline while younger flat-lying sediments of the Tabuk plateau overlie them unconformably.   Radiometric dating of tuffs or pumice cobbles in this sequence will indicate the time span during which the folding took place.

Geomorphology

The central Cagayan Valley is about 40 km wide and con-tains three major rivers. The largest, the Cagayan River, is a large meandering stream which drains the southern Cagayan Valley and flows to the sea in the north.   The other two rivers, the Chico and Pinacanauan de Tuguegarao, are braided tributary streams.   The Chico River drains the Cordillera Central which rises to an elevation of 2,216 m west of the valley, and the Pinacanauan de Tuguegarao River drains the Sierra Madre which rises to 1,833 m to the east.

Four major geomorphologic areas may be recognized in the central Cagayan Valley. From west to east these are: 1) the Tabuk plateau, 2) the Cagayan anticlinal belt, 3) the Cagayan River plain, and 4) the homoclinal belt.

The Tabuk plateau is a Pleistocene alluvial fan and terrace complex at elevations between 170 and 30 m along the mountain front. It extends for 15 km between the mountain front and Pangul Anticline and is underlain by a complex of pyroclastic flows and fluvial conglomerates and sandstones which unconformably overlie folded Miocene and Pliocene strata. The alluvial fan, which has a one-degree slope along the southern part of the plateau, has been dissected by the Chico River in the north. Here a sequence of four major rock-cut terraces are well-preserved between Pangul Anticline and the mountain front. Following folding of the anticlines, the east-flowing Chico River was diverted to the north as a result of ponding by Pangul Anticline. At present there is a sharp, right-angle bend in the Chico River Valley where it changes from a consequent to a subsequent valley.

The Cagayan anticlinal belt is characterized by dissected anticlines and synclines of varying relief.   They are surficially expressed as linear trends of grass-covered rugged hills and ridges or escarpments separated by broad, flat, alluviated synclinal valleys which constitute good grazing land and areas for rice production (Durkee and Pederson, 1961). Pangul Anticline, a large breached anticline with 350 m of relief, has the greatest relief of any of the anticlines in the area. In contrast, Cabalwan Anticline has the lowest relief, about 100 m. Wasson and Cochrane (1979) suggest that dissection of the anticlinal hills is not far advanced. They note that headward extension of the drainage lines has reached the axes of all anticlines. These drainage lines are primarily consequent, but important subsequent valleys do occur. Homoclinal ridges, upheld by resistant pyroclastic flow deposits, conglomerates, and sandstones, have been formed by the subsequent streams and are extensive along the flanks of the anticlines.

Several different erosional processes are operating on the flanks of the anticlines. Slope wash and soil creep are the two most important processes (Wasson and Cochrane, 1979). Surficial layers are being transported down slopes as indicated by imbricated pebble and cobble bands as much as a meter beneath the surface at several localities. Slumping also occurs and is most extensive in interbedded thin sandstones and mudstones of the Lower Member of the Ilagan Formation exposed in Pangul Anticline.

Many of the National Museum archaeological sites and fossil localities in the anticlinal belt occur in or are associated with 10 to 80 cm thick discontinous gravel beds previously thought to be terrace deposits (Lopez, 1972). Detailed stratigraphical studies of the anticlines indicate that the gravels are lag gravels derived from a conglomerate or conglomeritic sequence in the upper part of the Awidon Mesa Formation. Along the northern part of Cabalwan Anticline, a massive 5.5 m thick pebble-cobble conglomerate is present within the Awidon Mesa Formation. At the southern part of the anticline, stripped (terrace-like) surfaces are developed on this conglomerate which is more resistant to erosion than the surrounding finer-grained sediments. Along the southern nose of the Anticline, erosion has been more extensive resulting in lag gravel-covered hills.

The Cagayan River plain is a flat, synclinal, alluviated plain 10 to 20 m above sea level between the anticlinal and homoclinal belts. The Cagayan River meanders across the plain eroding along cutbanks and depositing sediments on large point bars, such as the bar just southwest of Tuguegarao which is up to 1.5 km in radius. Several terrace levels are well-developed along the eastern side of the Cagayan Valley. Durkee and Pederson (1961) have mapped the

Isabela Syncline along the same general trend as the river, and Wasson and Cochrane (1979) have noted that the river is, therefore, a consequent stream.

Two major cuestas form the homoclinal belt along the east side of the Cagayan Valley. These are upheld by the Callao Formation and the Lower Member of the Ilagan Formation. The Ilagan Formation which dips to the west at about 10° forms a steep east-facing, north-south trending escarpment with 120° m of relief along the Pinacanauan de Tuguegarao River. The river flows east of the escarpment in a strike valley developed in the underlying, less resistant shales of the Baliwag Formation before it cuts through the cuesta at Penablanca and joins the Cagayan River. Slumps are common along the escarpment which consists of thinly bedded sandstones and mudstones. Farther to the east, the Callao Limestone forms the most prominent cuesta and escarpment with up to 400 m of relief. Karst topography is well-developed on the 10° dip slope where archaeologists of the National Museum have located numerous cave sites. The geomorphology of these sites has recently been briefly described by Wasson and Cochrane (1979). One major stream, the Pinacanauan de Tuguegarao River, cuts across the Callao cuesta forming the Callao canyon. This river, which parallels the strike for a short distance between the two ridges, has been referred to as a consequent stream by Wasson and Cochrane (1979).

Paleontology

The first mammal fossils recovered from the Pleistocene strata of the Cagayan Valley were rhinoceros teeth. The teeth, found in 1936 near Tabuk, were described by von Koenigswald (1956) who noted that they were not identical with the species from China, Taiwan, or Indonesia. He assigned the teeth to a new species, Rhinoceros philippinensis nov. sp., and suggested that it was an endemic species.

In the early 1970's, poorly preserved remains of a Middle Pleistocene vertebrate fauna were found in the central Cagayan Valley during archaeological investigations (Fox, 1971; Fox and Peralta, 1974). Teeth are well-represented in the fossil collections and allow the preliminary identification of the elephants: Stegodon and Elephas, and rhinoceros, carabao (Bovidae), pig, and crocodile. The recovery of antlers and fragments of carapace indicate the presence of deer and turtle. All the finds so far are disarticulated, and the larger fossils are abraded indicating they were transported by streams before burial. From a brief description of the morphology of the elephant teeth, Lopez (1971) suggested that two forms of Stegodon were represented by the finds near Solana, Cagayan. Due to a lack of com-

parative material, he did not assign them to a particular
species.   Maglio (personal communication with Dr. Richard
Shutler, Jr., project archaeologist, 1976) examined one of
the Elephas molars from Solana and classified it as Elephas
cf. maximus but noted that it has a similar structure to the
now extinct Elephas hysudrindicus.

During the 1978 and 1979 field seasons, 32 vertebrate
localities containing primarily teeth and bone fragments,
were found along both flanks of Cabalwan, the northwest flank
of Enrile and the northeast flank of Pangul anticlines.
Although most were found as lag on the Awidon Mesa Formation,
5 localities were discovered where fossils occur in situ in
the upper part of the Awidon Mesa.

The most significant discovery was a small in situ
elephant skull, probably Stegodon, with two complete upper
molars.   This was found in an indurated sandstone along the
northeast flank of Pangul Anticline.   A number of bone
fragments and a worn pig tooth were recovered on the surface
in the area which can be radiometrically dated since tuffs
occur a few meters above and below the elephant skull.   Stone
tools occur on the surface at this site but none, as of yet,
have been recovered in situ.   One cobble tool, composed of
basalt, which was found on the surface appears to be ancient
because all surfaces are differentially weathered.

In addition to the elephant locality, four other fossil
sites were discovered where vertebrates occur in situ.   At
one locality along the western flank of Cabalwan Anticline,
carabao (Bovidae) teeth, vertebrae, and canon bones were
scattered on the surface and traced to a sandstone outcrop
where a tooth and several bones were recovered in situ.   The
fossils occurred in cobble-size clay balls in the basal
large-scale trough cross-bedded sandstone of a point bar
sequence that is overlain by a radiometrically datable pumice
conglomerate.

Plant fossils are present in the Ilagan Formation as well
as the Awidon Mesa.   Tree molds and permineralized wood are
common along with leaf impressions.   Fern leaves are particularly well-preserved and abundant in some tuffs.

Pleistocene vertebrate fossils have been found throughout
the Philippines (Beyer, 1956; von Koenigswald, 1956) indicating that land bridges once connected the Philippines with
mainland Asia.   It has been suggested that land bridges once
existed between Luzon, Taiwan, and Mainland China and between
the southern Philippines and the islands of Indonesia on the
Sunda shelf (Fig. 10) (DeTerra, 1943; Movius, 1949; von
Koenigswald, 1956; Sartono, 1973).   Land bridges must have

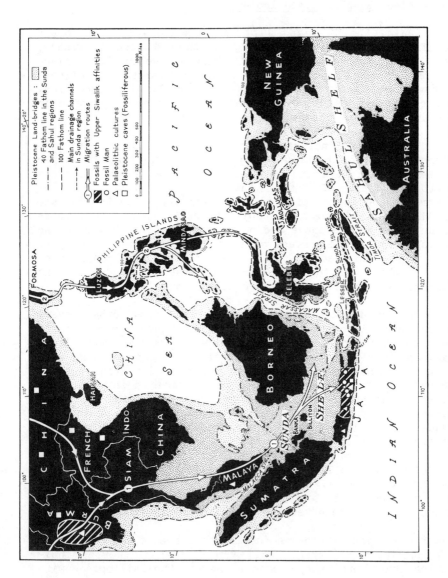

Figure 10. Pleistocene land-bridges connecting the Philippines and Asia (from Movius, 1949). Reprinted
by permission.

existed at those times when sea level dropped during the Pleistocene, thus allowing large vertebrates to migrate to the Philippines. The exact migration route cannot be stated with certainty at this time because tectonic activity has broken any possible bridge. Von Koenigswald (1956) favors the hypothesis that a Pleistocene land bridge existed between Luzon and Taiwan, noting that the fossil fauna of Taiwan contains virtually the same elements as that of the Philippines and that certain elements of the modern fauna are similar. Fox and Peralta (1974) also suggest that a bridge existed between Luzon and Taiwan considering that the mountains east, south, and west of the valley would have formed barriers to migration from the south. The possible land bridge between Luzon and Taiwan is formed by a submerged ridge, the North Luzon Ridge (Mammerickx et al., 1976), which now has a short, 2,000 m deep gap in it (Fig. 11). Considering the active Pleistocene tectonics of the region, faults such as those mapped on the ridge (Fig. 2) by Karig (1973) may easily have destroyed any Pleistocene land bridge. Further studies of the newly recovered vertebrate fossils may provide the necessary evidence to conclusively locate the Pleistocene land bridge or land bridges.

Facies and Environments of Deposition

The Plio-Pleistocene sediments of the central Cagayan Valley are divided into five major lithofacies. The facies and respective depositional environments as inferred from sedimentary structures are (1) the interbedded fine-grained sandstone and mudstone facies--delta front distal bar and distributary mouth bar, (2) the lenticular trough cross-bedded, medium-grained sandstone and siltstone facies--delta plain distributary channel, levee, and flood basin, (3) the polymictic conglomerate, trough cross-bedded sandstone and claystone facies--low energy fluvial channel and flood plain, (4) the polymictic clast-supported conglomerate and sandstone facies--high energy channel bar and gravel sheet, and (5) the massive matrix-supported pebble to boulder conglomerate and pyroclastic sandstone facies--ignimbrites and lahars, including both mudflows and debris flows.

The vertical sequence and lateral distribution of facies indicate a regression of the Pliocene sea as the basin filled with detritus from the surrounding volcanic arcs. The first three facies mentioned above occur in a vertical sequence throughout the central Cagayan Valley and document the transition from deltaic to fluvial sedimentation during the Pliocene. The last two facies interfinger with the fluvial polymictic conglomerate, trough cross-bedded sandstone and claystone facies along the western side of the Valley and represent an alluvial fan which formed at the base of the

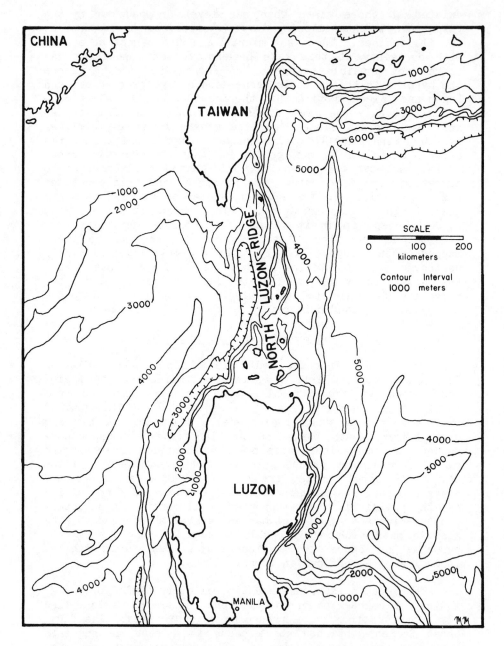

Figure 11.   Bathymetric map of the Luzon-Taiwan region (after
Mammerickx et al., 1976)

Cordillera Central as a result of Pleistocene tectonic and volcanic activity.

The Interbedded Fine-Grained Sandstone and Mudstone Facies

This facies is the dominant facies of the Lower Member of the Ilagan Formation and is well exposed along the Pinacanauan de Tuguegarao River. It reaches a thickness of 120 m where it overlies the prodelta claystones of the Baliwag Formation and forms a resistant escarpment along the river. The facies is also exposed at Pangul Anticline where it overlies prodelta claystones of the Buluan Formation of the Mabaca River Group. In the upper part of the Lower Member of the Ilagan Formation, the facies interfingers with and is overlain by the lenticular trough, cross-bedded, medium-grained sandstone and siltstone facies.

The interbedded fine-grained sandstone and mudstone facies is predominantly composed of very thin to thin, inter- bedded dark greenish gray (5 GY 4/1) to yellowish orange (10 YR 8/6) well-sorted sandstones and mudstones with subordinate siltstones. The sandstones and mudstones which usually ex- hibit parallel laminations or flaser and lenticular bedding are of constant thickness and laterally continous for at least several meters. The interbedded sandstone and mudstone sequences are commonly up to several meters thick. More massive appearing sandstones up to several meters thick are common in the facies. The sandstones, which are also of the uniform thickness and laterally continous, contain flaser bedding and occasional small-scale trough cross-bedding. Several mudstones up to 3 m thick are present in the facies. Lenticular bedding with connected and single lenses up to 2 cm thick is common in these mudstones. The facies contains a variety of fossils. Plant fragments are abundant along many bedding planes, while burrows up to 2.5 cm in diameter are common throughout the facies. Fractured or abraded pelecy- pods and gastropods are concentrated in some sandstones along with shell molds. They may be concentrated along bedding planes or be distributed throughout the sandstones in varying orientations. Shark teeth also occur in the facies but are rare. Calcareous and ferruginous, disk-to blade-shaped concretions up to boulder size also occur along many of the horizons that are fossiliferous.

Because of the lithology, thickness and lateral con- tinuity of the beds, sedimentary structures, and fossil content, the interbedded fine-grained sandstone and mudstone facies is interpreted as delta front distal bar and distribu- tary mouth bar deposits. The thin interbedded sandstones and mudstones with parallel laminations, lenticular bedding, and flaser structures are distal bar deposits similar to those described by Coleman and Gagliano (1965), Donaldson, Martin, and Kanes (1970), and Miall (1979). The thicker sandstones may exhibit small-scale trough cross-bedding and often contain fractured and abraded pelecypods and gastropods. These are interpreted as distributary mouth bar deposits. The thickness of the delta front distal bar and distributary mouth bar deposits (120 m) suggests that they were deposited by a river-dominated delta. River-dominated deltas are characterized by thick delta front deposits which accumulate as the delta advances by the continuous addition of sediment to the delta front by distributary channels (Miall, 1979). Thick delta front deposits do not form in deltas that are dominated by waves or tides which tend to deposit more of the sediment in offshore tidal sand ridges or shoreface deposits (Miall, 1979).

The Lenticular Trough Cross-Bedded Medium-Grained Sandstone and Siltstone Facies

The lenticular trough cross-bedded medium-grained sandstone and siltstone facies occurs in the upper part of the Lower Member of the Ilagan Formation where it interfingers with and overlies the interbedded fine-grained sandstone and mudstone facies. The only exposures occur at the Pangul Anticline and along the upper part of the escarpment along the Pinacanauan de Tuguegaro River where the maximum thickness measured is 24 m. The preserved thickness of this facies may be greater at Pangul Anticline, but abundant slumping and vegetative cover preclude accurate measurement.

The facies is composed of massive, lenticular, medium-to-coarse-grained, yellowish gray (5 Y 7/2) sandstones up to 9 m thick which cut into finer grained deposits dominated by thin parallel laminated siltstones and mudstones. Occasional thin very fine-grained sandstones with flaser structures also occur associated with the siltstones and mudstones. The lenticular sandstones have scoured bases and often display large- and small-scale trough cross-bed sets up to 50 cm thick. Some of the lenticular sandstones exhibit channel-fill cross-bedding (Reineck and Singh, 1975). The associated finer grained deposits are commonly bioturbated and contain plant fragments.

The lenticular trough cross-bedded medium-grained sandstone and siltstone facies is interpreted to represent delta plain distributary channel, levee, and associated flood basin deposits. Included are subaqueous distributary channels and levees, but these were not differentiated. Characteristics of lenticular, cross-bedded, distributary channel sandstones with scoured basal surfaces have been described by many authors and are summarized by Coleman and

Gagliano (1965) and Wright (1979), who have also described
the characteristics of delta plain levee and flood basin
deposits which include marsh, swamp, interdistributary bay,
and mudflat deposits. These consist of thin sands, silts,
and clays that are bioturbated and often contain large
amounts of organic material which may form peat or coal.
While no peat or coal deposits were found in outcrops of this
facies in the Cagayan Valley examined by the authors, they
have been reported to occur in the Ilagan Formation by Corby,
et al. (1951). In addition to the evidence from the rock
types and sedimentary structures, the stratigraphic
occurrence of the facies supports the delta plain
environmental interpretation. The facies interfingers with
and overlies delta front deposits and interfingers with
fluvial conglomerates and sandstones of the polymictic
conglomerate, trough cross-bedded sandstone, and claystone
facies.

The Polymictic Conglomerate, Trough Cross-Bedded Sandstone, and Claystone Facies

This facies is the most extensive of the
Plio-Pleistocene deposits. It is exposed along Enrile and
Pangul anticlines where it forms the Upper Member of the
Ilagan Formation and at the latter, reaches a thickness of
570 m. This facies also occurs in the Awidon Mesa Formation
where it is interbedded with the polymictic clast-supported
conglomerate and sandstone facies and the massive matrix-
supported pebble-to-boulder conglomerate and pyroclastic
sandstone facies. In the northern part of Cabalwan
Anticline, it reaches a thickness of 120 m.

The polymictic conglomerate, trough cross-bedded
sandstone and claystone facies are composed of pale yellowish
orange (10 Y 8/6) granule to cobble conglomerates, coarse to
fine-grained sandstones and siltstones, and pale olive (10 Y
6/2) claystones which characteristically occur in fining-
upward sequences 5 to 40 m thick. The Pliocene conglomerates
of this facies are thin channel lag conglomerates composed
primarily of granules and pebbles at the base of thick
sandstones. The Plio-Pleistocene conglomerates and conglomeratic sandstones are thicker, to 7 m, and contain coarser
pebble- to cobble-size clasts. These are dominantly disk to
equant in shape, subrounded to rounded, and composed of
basalt and porphyritic andesite with minor amounts of quart-
zite, chert, and sedimentary rock fragments. The conglomerates have sharp but irregular scoured basal contacts and
commonly display large-scale trough cross-bedding in sets up
to 1 m thick. They are overlain by trough cross-bedded
sandstones up to 12 m thick which often fine upward from very
coarse to very fine sand. The sandstones exhibit large-scale
trough cross-beds up to 1 m thick which grade upward to

small-scale trough cross-beds. Valid paleocurrent estimates
were not obtainable due to limited exposures and the highly
variable directional nature of the trough cross-bedding.
Horizontal stratified sands which are laterally continuous
for many meters occasionally occur at various levels in the
upper part of the sandstones. The fining-upward sequences
are capped by thin parallel laminate and cross laminated
silt-stones and massive claystones which are commonly up to
20 m thick but with a maximum thickness of 35 m.   The
claystones, which may be sandy or contain thin sand lenses,
vary from poorly exposed to covered.   Numerous claystones
contain horizons with blocky structure or iron rich pisolites
which are interpreted as paleosols.

    The conglomerates and sandstones of the polymictic
conglomerate, trough cross-bedded sandstone, and claystone
facies form resistant, laterally extensive tabular deposits
of nearly uniform thickness. They are commonly traceable in
scarp faces of hogbacks along the flanks of anticlines for at
least several kilometers. However, lenticular channel deposits and numerous thin, lenticular tuffs and thicker, tabular
tuffaceous intervals up to 10 m thick occur within this
sequence. Permineralized wood is commonly found in sandstones, siltstones, and claystones associated with the tuffaceous intervals. In addition, leaf impression, primarily
ferns, are common in many of the tuffs, and vertebrate
fossils have been found in situ in conglomerates, sandstones
and claystones of the facies in the Awidon Mesa Formation.

    The polymictic conglomerate, trough cross-bedded
sandstone, and claystone facies are interpreted to have been
deposited by meandering streams.   The conglomerates and
sandstones are lateral accretion deposits which accumulated
as point bars of intermediate to high sinuosity streams. The
major characteristics of lateral accretion or point bar
deposits have been well-documented by numerous investigations
of modern and ancient streams (Allen, 1965, 1970; Visher,
1965; McGowan and Garner, 1970; Walker and Cant, 1979). The
conglomerates represent channel lag in the bottom of
migrating channels and were overlain by trough cross-bedded
sand as the river migrated laterally by eroding the concave
bank. As described by Harms and Fahnestock (1965), the flow
phenomena range from the upper part of the lower flow regime
to the lower part of the lower flow regime. The horizontally
stratified sands which occasionally occur in the upper part
of some sandstones indicate flow over a plane bed at high
velocities in the upper flow regime of Harms and Fahnestock
(1965).   The parallel and cross-laminated siltstones and
massive claystones represent vertical accretion deposits outside the main channel.   The silts and clays settled from
suspension, thus building up the flood plain.   The thin len-

ticular tuffs which occur in the facies associated with the siltstones and claystones were deposited on the upper part of the point bar or in swales on the flood plain on which soils developed as indicated by pisolitic paleosols and horizons with blocky structure.

The geometry of the sandstones and claystones provides further evidence regarding the nature of the stream system. The sandstones are laterally continuous, hundreds of meters to several kilometers in length, and fairly uniform in thickness. Modern meandering streams deposit similar sheet-like sand bodies as they migrate across flood plains (McGowan and Garner, 1970). Moody-Stuart (1966) has suggested that sheet-like sandstone geometry is a diagnostic characteristic of meandering stream channel sands. The sandstones are thickest, 6-12m, in the Upper Member of the Ilagan Formation and thinner, 2-6 m, in the Awidon Mesa Formation. The Ilagan Formation sandstones were deposited by a large meandering stream with a channel up to 12 m deep. This stream flowed to the north along the axis of the valley. The thinner Awidon Mesa Formation sandstones represent smaller meandering tributary streams which flowed along the western side of the valley. Ratios of thickness between lateral accretion and vertical accretion deposits have been used to try to interpret characteristics of the stream systems during deposition. In this facies the ratios range between 1:2 and 1:3 as the vertical accretion deposits are 2 to 3 times thicker than the lateral accretion deposits. Instead of providing further evidence on the nature of the stream system, these ratios most probably reflect the width of the alluvial plain and rates of subsidence and orogenesis as noted by Collinson (1977) and Friend (1977).

Recent research on meandering stream deposits has indicated that they may be more complex than previously thought. Jackson (1977) notes that widely used criteria for meandering streams do not occur in all meandering streams or they can exist in non-meandering streams. Epsilon cross-stratification for example, which Moody-Stuart (1966) considers a diagnostic characteristic of laterally accreted meandering stream deposits, has not been observed in the Cagayan Valley. Observations by Allen (1970) and Jackson (1977) indicate that epsilon cross-stratification is rare in many modern and ancient deposits.

In summary, the interpretation of the polymictic conglomerate, trough cross-bedded sandstone, and claystone facies is based on the fining-upward grain size, sedimentary structure sequences, geometry of the sand bodies and thick flood plain deposits, and stratigraphic position of the facies between deltaic and braided stream deposits.

The Clast-Supported Polymictic Conglomerate and Sandstone Facies

Conglomerates and sandstones of this facies occur throughout the Awidon Mesa Formation but are thicker, up to 50 m thick, and more numerous toward the southern and western part of the central Cagayan Valley. The clast-supported polymictic conglomerate and sandstone facies is well exposed along the flanks of the anticlines where it overlies and interfingers with the polymictic conglomerate, sandstone, and claystone facies and the massive matrix supported conglomerate and pyroclastic sandstone facies.

A variety of clast-supported conglomerates and fine- to coarse-grained sandstones occur in this facies which could be assigned to numerous alluvial gravel facies as outlined by Rust (1979). Massive conglomerates up to 22 m thick are most common in the southern part of the central Cagayan Valley. Low angle large-scale trough cross-bedded conglomerates which grade upward to sandstone within each cross-bed set occur throughout the eastern and northern part of the study area where they may attain a thickness of 6 m. Convolute bedding is well developed in some of the sandstones in these sequences. Planar cross-bedded and horizontally stratified conglomerates and sandstones occur in the facies but are much less common. The horizontally stratified conglomerates and sandstones reach 2 m in thickness while the planar cross-bed sets are not thicker than 30 cm. The conglomerates and sandstones just described occur in a vertically unordered sequence characterized by abrupt textural changes.

The clast-supported polymictic conglomerate and sandstone facies is interpreted to have been deposited by eastward-flowing braided streams as part of an extensive Pleistocene alluvial fan complex. The imbricated horizontally bedded to massive clast-supported conglomerates were deposited as gravel sheets and low relief longitudinal or diagonal bars (Smith, 1974; Hein and Walker, 1977; Rust, 1978). The large-scale trough cross-bedded conglomerates which grade upward to sandstones and the planar cross-bedded conglomerates and sandstones are interpreted to be more distal channel bar deposits. Rust (1979) has described similar large-scale trough cross-bedded conglomerates which fine upward as a response to channel migration or shallowing of the water over bars and channels as they were accreted in the distal reaches of a braided stream. The planar cross-bedded conglomerates and sandstones were formed by downstream migration of transverse bars (Smith, 1970; Hein and Walker, 1977). The abrupt and vertically unordered textural changes in the facies suggest that there was frequent shifting of shallow channels and bars (Harms et al., 1975).

Along Cabalwan Anticline disarticulated vertebrate fossils have been found in situ in these deposits. Toward the southern end of the Anticline, laterally equivalent sediments have been extensively eroded leaving lag gravels containing additional vertebrate fossils and tektites.

The Massive Matrix-Supported Pebble-to-Boulder Conglomerate and Pyroclastic Sandstone Facies

The massive matrix-supported pebble-to-boulder conglomerate and pyroclastic sandstone facies occurs throughout the Awidon Mesa Formation increasing in thickness in the upper part and along the western portion of the central Cagayan Valley. It is particularly well-exposed along the Tabuk plateau where it is up to 25 m thick. The facies thins to 2 to 3 m along the flanks of Cabalwan and Pangul anticlines where it interfingers with the polymictic clast-supported conglomerate and sandstone facies and the polymictic conglomerate trough cross-bedded sandstone, and claystone facies. However, along the western flank of Enrile Anticline, thicker deposits occur which are up to 16 m thick. The facies extends discontinuously into the Cordillera Central where a thickness of 300 m has been recorded by Durkee and Pederson (1961) at Awidon Mesa.

The massive matrix-supported pebble-to-boulder conglomerate and pyroclastic sandstone facies is composed of very light gray (N8) to light gray (N7) massive pyroclastic flow which includes oligomictic and polymictic pebble-to-boulder conglomerates, breccias, and sandstones of dacitic composition. The matrix of the conglomerates consists of medium- to coarse-grained sand of dacitic composition. The oligomictic conglomerates are composed of either rounded, disk- to equant-shaped pumice clasts or subangular to angular dacite or andesite clasts while the polymictic conglomerates contain subrounded basalt clasts in addition to dacite or pumice. The basalt clasts are usually only pebble to cobble size in contrast to dacite and pumice clasts which may reach boulder size. The massive pyroclastic flow units, up to 5 m thick, may be either normally or reversely graded. Small or large scale low angle trough cross-bedded sandstones or conglomeritic sandstones occur at the base and occasionally at the top of the deposits along with thin tuffs. The orientation of clasts varies within the facies from clasts oriented parallel to the bedding or imbricated to randomly oriented clasts. Some deposits have sharp irregular basal contacts indicating that the underlying deposits were scoured by the flow while other deposits have planar contacts suggesting that scouring did not occur. The geometry of the facies varies with age of the deposit. Older deposits are lenticular or grade laterally into the clast supported poly-

mictic conglomerate and sandstone facies. The younger deposits in contrast, occur as large sheets which are traceable
throughout most of the study area. These younger deposits
are especially valuable as marker beds.

The massive matrix-supported pebble-to-boulder conglomerate and pyroclastic sandstone facies is a complex of non-
welded or sillar ignimbrites and lahars. Some deposits
represent individual ignimbrites or lahars, but the facies
commonly occurs as a complex in which several ignimbrites may
occur in a sequence, or individual ignimbrites may be interbedded with lahars or grade upward to lahars. In many deposits up to six distinct flow units can be recognized.

The term ignimbrite is used as defined by Sparks et al.
(1973) and Sparks (1976) to refer to deposits of Pelean
eruptions which are characterized by three distinct layers in
a vertical sequence. These are (1) a lower ground surge
deposit, (2) the main ignimbrite flow unit, and (3) a fine
ash deposit. The low angle cross-bedded deposits of the
massive matrix-supported conglomerate and pyroclastic
sandstone facies are usually between several centimeters and
1 m thick and are interpreted to be ground surge deposits
(Sparks and Walker, 1973) which preceded the main flow. In
deposits with many flow units, some of the surge layers may
have been deposited at the top of pyroclastic flow sequences
as recently described by Fisher (1979). The main flow units
of the Cagayan Valley ignimbrites are massive matrix-
supported pebble to boulder breccias up to 5 m thick. They
have a diagnostic crystal rich basal layer (Walker, 1971;
Sparks et al., 1973) which is finer than the rest of the unit
and which is characterized by a normal grading of lithic
clasts and a reverse grading of large pumice clasts. "Fossil
fumaroles", vertical pipe-like structures which lack the
fine-grained material of the deposit (Walker, 1971), are preserved in many deposits and were formed as gas escaped from
the settling flow. The fine ash deposit or coignimbrite ash
(Sparks and Walker, 1977) is a thin, very fine ash deposit
which, in a complete sequence, overlies the main ignimbrite
flow unit.

The ignimbrites indicate that both Plinian and Pelean
eruptions occurred during the Pleistocene and probably in the
Cordillera Central which is the closest possible source.
Valley-filling pyroclastic deposits occur in the Cordillera
at Awidon Mesa and several mesa-like localitites further to
the west. Short duration, highly explosive Plinian eruptions
preceded the Pelean eruptions. This is indicated by the
abundance of pumice clasts in the ignimbrites which typically
originate during an initial Plinian phase of volcanism which

produces pumice fall deposits (Sparks et al., 1973). The ignimbrites were generated by subsequent Pelean eruptions of dacitic magma which likely occurred during a period of a few years (Williams and McBirney, 1979). The layers in the ignimbrite deposits were formed by differentiation of the material that was erupted (Fisher, 1979). The ground surge was formed by collapse of the outer part of the eruption column. Progressive collapse of the eruption column interior produced the voluminous high concentration pyroclastic flows which, according to Sparks (1976), are commonly laminar in their movement. The coignimbrite ash was deposited from an ash cloud that was segregated from the surface of the flow. According to Fisher (1979), it may be overlain by a thin fallout deposit with characteristics of a surge deposit.

Lahars are common in the massive matrix-supported pebble-to-boulder conglomerate and pyroclastic sandstone facies but are not as numerous as ignimbrites. Lahar is an Indonesian word for volcanic breccias which were transported down the slopes of a volcano by water (van Bemmelen, 1949). As suggested by Crandell (1971) and Williams and McBirney (1979), lahar is here used as a general term for both volcanic mudflows and debris flows. It is pertinent to note, however, that the massive matrix-supported pebble to boulder conglomerate and pyroclastic sandstone facies contains both debris flows and mudflows. Mudflows, which contain at least 50% sand, silt, and clay (Varnes, 1958), are the most common type of lahar, while debris flows, which contain less than 50% sand, silt, and clay, are restricted to the western side of the study area.

As Crandell (1971) notes, there is no single feature which may be used to distinguish lahars from all other kinds of coarse deposits. Some lahars are normally graded (Mullineaux and Crandell, 1962) while others have been described as reversely graded (Schmincke, 1967). Considering flow dynamics, Hampton (1975) suggested that debris flows may be normally or reversely graded. Clast orientation varies between lahars as clasts may be oriented parallel to the bedding (Fisher, 1971; Enos, 1976) or lack preferred orientation (Reineck and Singh, 1975). The character of the basal contact also varies as the base of the underlying deposit may be scoured (Schminke, 1967) or lack scour features (Enos, 1976). Despite these variations, which also occur in the Cagayan Valley lahars, there are several distinguishing features which make it possible to identify the lahars. These are the massive, poorly sorted, matrix-supported character of the deposits and the occurrence of subrounded to subangular clasts of various compositions. The clast composition and rounding enables lahars to be differentiated from ignimbrites as ignimbrites contain only volcanic clasts which are dominantly subangular. Lahars also lack the associated ground surge deposit, the upper fine ash deposit, and the "fossil fumaroles" which are diagnostic of ignimbrites.

The lahars originated as pyroclastic flows which were mobilized by water from rains during or after volcanic eruptions. The water-soaked pyroclastic debris flowed from the Cordillera Central along drainage ways and then spread out into the Cagayan Valley along stream channels and floodplains. Accretionary lapilli, which commonly form in ash clouds when it rains, have not as of yet been found in the Cagayan Valley pyroclastic deposits.

The stratigraphic occurrence and geometry of the ignimbrites and lahars are illustrated in Figure 6. The older discontinuous lenticular deposits which help define the base of the Awidon Mesa Formation are interpreted as distal ignimbrites and lahars. These grade laterally to cross-bedded dacitic conglomerates and sandstones within several hundred meters. The younger extensive sheet-like deposits are interpreted as more proximal ignimbrites and lahars.

The massive matrix-supported pebble-to-boulder conglomerate and pyroclastic sandstone facies and clast-supported polymictic conglomerate and sandstone facies form a Pleistocene alluvial fan complex. The two facies comprise a coarsening-upward sequence of stream deposited conglomerates, ignimbrites, and lahars which reflect the continuing uplift of the Cordillera Central and intensified volcanic activity. The coarsening-upward reflects this uplift as the fan complex prograded during the Pleistocene and buried the finer distal deposits by coarser proximal detritus. Similar coarsening-upward sequences in Norway were described and attributed to tectonic activity by Steele and others (1977).

Paleogeography and Paleoenvironments

The Cagayan basin began to form in the Late Oligocene to Early Miocene epochs following the polarity reversal of the Luzon arc system (Fig. 3) and the initial uplift of the ancestral Cordillera Central (Durkee and Pederson, 1961; Karig, 1973). The Early Miocene basin had a northeasterly trend similar to other Philippine structures such as Palawan Island (Christian, 1964). Reef limestones of the Early Miocene Sicalao Formation accumulated along the western periphery of the basin while quartz-deficient clastic marine sediments of the Gatangan Creek Formation and Mabaca River Group were deposited in the rest of the basin. During the Middle and Late Miocene, bathyl turbidites of the Gatangan Creek Formation and Mabaca River Group were deposited as the

basin subsided.   Localized uplift, warping, and perhaps
faulting, brought about a significant change in basin morphology during the Miocene as the basin axis became oriented
north-south and migrated to the east (Fig. 4), and the
eastern shelf of the basin became exposed resulting in the
erosion of thousands of feet of Miocene sediments (Christian,
1964).   A transgressive Late Miocene sea then overlapped the
eastern shelf depositing reef limestones of the Callao
Formation.   During the Late Miocene, prodelta clays of the
Baliwag Formation were deposited along the eastern margin of
the basin, and in the central and western portions, clays of
the Late Miocene Buluan Formation of the Mabaca River Group
were deposited.

Paleogeographic reconstructions and paleoenvironmental
interpretations of the central Cagayan Valley terrestrial
sequence are based on the recognition of facies and their
distribution.   The vertical sequence and lateral distribution
of the Plio-Pleistocene facies documents a regression of the
Pliocene sea as the basin filled with detritus from the
surrounding volcanic arcs.   An interpretive cross section of
the Plio-Pleistocene sediments before and after folding is
presented in Figure 12.   The dates used in the following
discussion   are   based   on   previous   investigations   as
radiometric dating of samples collected during this study has
not yet been completed.

The beginning of the Pliocene epoch is marked by a transition   from   marine   to   terrestrial   sedimentation   as
progressive uplift of the Cagayan region developed.   Lower
Ilagan delta front sediments of the interbedded fine-grained
sandstone and mudstone facies were deposited on Miocene prodelta clays of the underlying Baliwag and Buluan Formations.
Thinly interbedded fine-grained sandstones, siltstones, and
mudstones formed a distal bar on the seaward margin of the
advancing delta front.   Some of the deformed bedding observed
in the thinly interbedded sandstones and siltstones likely
represents penecontemporaneous slumping as the delta subsided
and advanced.   Shell remains, common in the sandstones,
suggest that they were transported by higher energy currents
which deposited the thin sand beds.   Small burrowing organ-
isms   lived   along   the   distal   bar   as   indicated   by   the
occurrence of burrows up to 2.5 cm in diameter.   Sediment was
transported to the delta front distributary mouth bar by
subaqueous   and   subareal   distributary   channels   of   the
lenticular, trough cross-bedded, medium-grained sandstone and
siltstone facies.   The distributary channels flowed across a
delta plain composed of finer grained sediments of the facies
which were deposited during storms and periods of flood.
These finer grained sediments were bioturbated to varying
degrees by plants and burrowing animals.   The delta front and

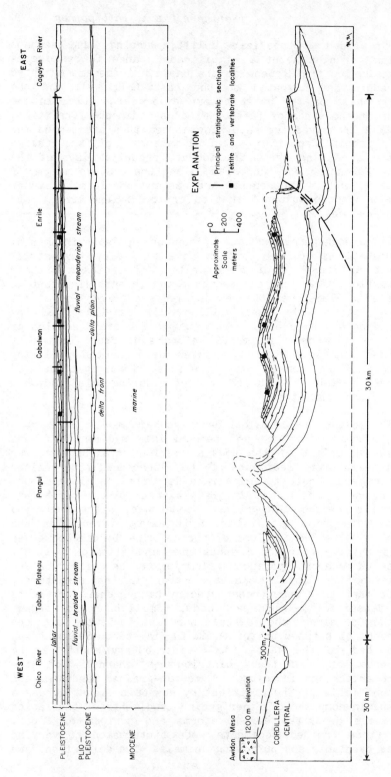

Figure 12. Interpretive cross sections of the central Cagayan Valley before and after Middle Pleistocene folding.

delta plain sediments of the Lower Member of the Ilagan
Formation were deposited by at least one river-dominated
delta. Wave and tidal influences did not have as great an
effect on delta growth as did the constant influx of sediment
by a river. Streams transported sediment to the delta from
three major areas, the infilled valley to the south and the
volcanic arcs to the east and west. At this time it cannot
be stated with certainty which sources were dominant during
the Pliocene.

At least one major meandering stream flowed to the north
across the central Cagayan Valley in the Plio-Pleistocene and
deposited 570 m of the Upper Ilagan polymictic conglomerate,
trough cross-bedded sandstone, and claystone facies. The
river flowed in a channel up to 12 m deep as indicated by the
thickness of the conglomerates and sandstones. Changes in
conglomerate thickness and clast size indicate the increasing
contribution of a westerly source as the basin filled and the
Cordillera Central was uplifted. The older conglomerates of
the Upper Member of the Ilagan Formation are thin and composed of granules and pebbles which were transported primarily from the southern part of the Cagayan Valley by a
major axial stream or from the Cordillera Central which was
then a distant westerly source. The conglomerates thicken
and coarsen to pebble-cobble conglomerates in the upper part
of the Upper Member reflecting the increasing contribution of
a more proximal westerly source, the Cordillera Central
volcanic arc, which was active during the Pliocene.
Volcaniclastic sediments were reworked by streams and deposited in association with sandstones, siltstones, and claystones in the swales of point bars and on the flood plains
where soils developed as indicated by the numerous horizons
with blocky structure and several iron rich pisolitic
paleosols. Permineralized logs up to several meters long
indicate that trees were eroded, transported, and deposited
by the stream. Root casts and leaf impressions in tuffs
indicate a variety of plant cover existed in the area.

During the Pleistocene, uplift of the Cordillera Central
and volcanic activity intensified, and an alluvial fan
complex formed along the mountain front. Coarse clastics of
the clast-supported polymictic conglomerate and sandstone
facies were transported to the valley by braided streams.
The coarser sediments were deposited on the inner fan in the
southwestern part of the study area. Finer sediment was
deposited by distal braided streams on the toe of the fan or
transported by small meandering streams to the axial
meandering stream which probably flowed along the eastern
side of the valley where paleo-Cagayan River terraces now
occur. As uplift of the Cordillera continued, proximal
braided stream deposits developed farther out in the basin

covering the distal deposits (Fig. 12). Ignimbrites and lahars flowed along topographic depressions from the Cordillera Central to the valley where they spread out on the alluvial fan and alluvial plain of the paleo-Cagayan River. Durkee and Pederson (1961) have suggested that three little known vents described by Alvir (1956) may have been the source of the pyroclastic deposits. Flat mesa-like areas west of Awidon Mesa which are distinctive on topographic maps of the Cordillera Central are likely upheld by pyroclastic flow deposits similar to those at Awidon Mesa. This suggests that the source was west of Awidon Mesa. The dacitic composition and quartz phenocrysts in the pyroclastic flow deposits indicate that volcanism became more silicic in the Pleistocene in contrast to the mafic to intermediate volcanism of the Miocene and Pliocene and suggests that the Cordillera volcanic arc has matured. Ragland, Stirewalt, and Newcomb (1976) describe three volcanic belts in western Luzon, a tholeiitic belt in the west, a calc-alkaline belt, and a high-K calc-alkaline belt in the east. The high-K calc-alkaline belt reflects the maturity of the arc as high-K calc-alkaline volcanism occurs only during the advanced stages of island arc evolution (Gill and Gorton, 1973).

A terrestrial vertebrate fauna migrated to the Cagayan Valley via at least one land bridge during the Middle Pleistocene. A land bridge between Luzon and Taiwan is suggested by faunal and geographic considerations but cannot be proved at this time. When the vertebrates arrived, the valley contained a large, north-flowing, meandering stream with an extensive flood plain and an alluvial fan complex forming along the Cordillera Central mountain front. All the vertebrate fossils found in situ were transported and buried in the coarse deposits or flood plains of small meandering streams or distal braided streams between the toe of the alluvial fan and the paleo-Cagayan River (Fig. 12).

Tektites of possible extraterrestrial or impact origin were deposited by distal braided streams and/or small meandering streams on the plain between the alluvial fan and the paleo-Cagayan River (Fig. 12). The limited stratigraphic occurrence of tektites and the association with Middle Pleistocene fauna suggest that they are Middle Pleistocene in age and probably correlative with the 700,000 year B.P. tektites of Java (von Koenigswald, 1967). Among other theories, Dietz and McHone (1976) have recently suggested that the 700,000 year B.P. tektites of the Southeast Asian strewnfield may be products of the meteoritic impact at El'gytgyn, Siberia.

Little can be inferred at this time regarding Pleistocene climates. The planned laboratory research should provide a basis for interpretations pertaining to the Cagayan Valley. An important regional synthesis of Pleistocene climatic variations in Southeast Asia has been compiled by Verstappen (1975). He relates changes in sea level, temperature and more importantly, rainfall and humidity changes to changes in vegetative cover, soil formation, and landform development (Table 3). This information is important to consider in cultural interpretations of Pleistocene archaeologic sites (assuming the Cagayan Valley tools will be found in situ) because early human culture reflects adaptations to environments. Based on Verstappen's (1975) summary and fieldwork, a tree savanna environment probably predominated in the Cagayan Valley during the Pleistocene while monsoon forests grew along the mountain front and streams. A tree savanna environment must have existed before the Middle Pleistocene folding as indicated by the abundance of large grazing mammal fossils. The present vegetation patterns have been modified by agricultural practices, but tree savanna and grasslands appear to be the dominant vegetation with monsoon forests along the mountain front and streams.

The Cagayan Valley anticlinal belt formed in the Middle to Late Pleistocene (Fig. 12). Accelerated uplift of the Cordillera Central formed a slope which made it possible for gravity-sliding and asymmetric-to-overturned folds to form. The resulting folds led to changes in the drainage system along the mountain front. The formation of Pangul Anticline restricted stream flow to the east as indicated by the series of rock-cut terraces along the Tabuk plateau which documents the progressive diversion of the paleo-Chico River to the north. The terrace levels probably reflect uplift of the Cordillera and renewed downcutting. When the Chico River diverted to the north, the alluvial fan complex to the south became inactive and was partially dissected by streams. Erosion of the anticlines has provided sediment which has been accumulating in the adjacent synclines since folding took place. Numerous flat-lying dacitic ignimbrites and lahars along the Tabuk plateau (Fig. 12) indicate that dacitic volcanism continued after the folding episode.

The date when hominids arrived in the Cagayan Valley and started making tools is dependent on further archaeological investigations. If tools or hominid remains can be found in situ in the folded sediments of the anticlines, a Middle Pleistocene age is suggested by the association with Middle Pleistocene vertebrates and tektites. More precise dates will be available when radiometric dating of tuffs and pumice cobbles by fission track and K-Ar techniques is completed. If human forms inhabited the Cagayan Valley during the Middle Pleistocene, gravel sources for tools were easily accessible along the meandering and distal braided streams. If the

Table 3. Summary of Pleistocene climatic interpretations for
Southeast Asia (from Verstappen, 1975).

1. Sea level dropped 100 m converting 3,000,000 km^3 of shal-
    low warm seas to land; seawater temperatures dropped 4-
    5°C.

2. 30% drop in rainfall below present values during glacial
    periods; distribution of rainfall different.

3. Drop in rainfall produced a more pronounced dry season
    which caused drought stress in vegetation and changes in
    soils, fauna, and geomorphological processes.

4. Botanical evidence suggests drier conditions have been
    the exception in the past; humid tropical conditions are
    thought to be more normal. Aridity was never so severe
    that steppe or grass savanna could become established.
    Monsoon forest and tree savanna were probably the charac-
    teristic vertebrate environments during the glacials.

5. Humidity increases may have been responsible for the
    extinction of the grazing mammals.

arrival of the hominids post-dates the folding, gravels would also have been easily found after the folding due to erosional processes which have produced the lag gravels on many of the anticlines. The age documentation of the hominids in the Cagayan Valley is now dependent on archaeological excavations to find tools or hominid fossils in situ.

Acknowledgments

This research was funded by National Science Foundation grant INT-7901802 and the Philippine National Museum. Sincere thanks are given to National Museum Director Dr. Godfredo Alcasid, Assistant Director Dr. Alfredo Evangelista, and Anthropology Curator Dr. Jesus Peralta for their efforts in financing and organizing the fieldwork. This study could not have been completed without the enthusiastic assistance of many National Museum geologists. Special thanks are extended to Yolando Senires, Geology Division Curator, Louis Omana, Roberto de Ocampo, Severino Pascual, Melchor Aguilera, Nestor Bondoc, and Lina Flor. The cooperation of all the National Museum employees and the generous hospitality and assistance of countless Filipinos throughout the Cagayan Valley are also greatly appreciated. We are also indebted to Dr. Basil Booth of Imperial College and Birkbeck College, University of London, who assisted in the field interpretation of the ignimbrites and lahars.

References

Allen, J. R. L., 1965, Sedimentation and paleogeography of the Old Red Sandstone of Anglesey, North Wales: Yorkshire Geological Society Proceedings, v. 35, p. 139-185.

Allen, J. R. L., 1970, Studies in fluviatile sedimentation: A comparison of fining upwards cyclothems, with special reference to coarse member composition and interpretation: Journal of Sedimentary Petrology, v. 40, p. 298-323.

Alvir, A. D., 1956, A Cluster of little known Philippine volcanoes: Proceedings Eighth Pacific Science Congress, v. 2, p. 205-206.

Barnes, V. E., and Barnes, M. A., 1973, Tektites: Stroudsburg, Penn., Dowden, Hutchinson and Ross, p. 445.

Beyer, H. O., 1956, New finds of fossil mammals from the Pleistocene strata of the Philippines: National Research Council of the Philippines Bulletin No. 41, p. 1-17.

304   *Vondra et al.*

Beyer, H. O., 1961, Philippine tektites: Quezon City, Philippines, University of the Philippines.

Black, Davidson, 1933, Fossil man in China: Peiping, Geological Survey of China.

Bowin, C., Lu, R. S., Lee, C., and Schouten, H., 1978, Plate convergence and accretion in the Taiwan-Luzon Region: American Association of Petroleum Geologists Bulletin, v. 62, p. 1645-1672.

Caagusan, N. L., 1978, Source material, compaction history and hydrocarbon occurrence in the Cagayan valley basin, Luzon, Philippines, in SEAPEX Program, Offshore Southeast Asia Conference, Singapore: Southeast Asia Petroleum Exploration Society.

Christian, Louis B., 1964, Post-oligocene tectonic history of the Cagayan Basin, Philippines: The Philippine Geologist, v. 18, p. 114-147.

Coleman, J. M., and Gagliano, S. M., 1965, Sedimentary structures Mississippi River deltaic plain, in Middleton, G. V., ed., Primary sedimentary structures and their hydrodynamic interpretation: Society of Economic Paleontologists and Mineralogists, Special Publication 12, p. 133-148.

Collinson, J. D., 1977, Vertical sequence and sand body shape in alluvial sequences, in Miall, A. D., ed., Fluvial sedimentology: Canadian Society of Petroleum Geologists Memoir 5, p. 577-586.

Corby, G. W. et al., 1951, Geology and oil possibilities of the Philippines: Republic of the Philippines Department of Agriculture and Natural Resources Technical Bulletin 21, 363 p.

Crandell, D. R., 1971, Post glacial lahars from Mount Ranier Volcano, Washington: U.S. Geological Survey Professional Paper 677, 75 p.

DeTerra, H., 1943, Pleistocene geology and early man in Java: Transactions of the American Philosophical Society, v. 32, p. 437-464.

Deitz, R. S., and McHone, J. F., 1976, El'gytgyn: Probably world's largest meteorite crater: Geology, v. 4, p. 391-392.

Donaldson, A. C., Martin, R. H., and Kanes, W. H., 1970, Holocene Guadalupe delta of Texas gulf coast, in Morgan, J. P., and Shaver, R. H., eds., Deltaic sedimentation, modern and ancient: Society of Economic Paleontologists and Mineralogists Special Publication 15, p. 107-137.

Durkee, E. F., and Pederson, S. L., 1961, Geology of Northern Luzon, Philippines: American Association of Petroleum Geologists Bulletin, v. 45, p. 137-168.

Enos, P., 1977, Flow regimes in debris flow: Sedimentology, v. 24, p. 133-142.

Fisher, R. V., 1971, Features of coarse grained high concentration fluids and their deposits: Journal of Sedimentary Petrology, v. 41, p. 916-927.

Fisher, R. V., 1979, Models for pyroclastic surges and pyroclastic flows: Journal of Volcanology and Geothermal Research, v. 6, p. 305-318.

Fox, R. B., 1971, Ancient man and Pleistocene fauna in Cagayan Valley, Northern Luzon, Philippines: A Progress Report of the Philippine National Museum, Manila, p. 1-32.

Fox, R. B., and Peralta, J. T., 1974, Preliminary report on the paleolithic archaeology of Cagayan Valley, Philippines, and the Calalwanian industry: Proceedings of the First Regional Seminar on Southeast Asian Prehistory and Archaeology, Manila, National Museum, p. 100-147.

Friend, P. F., 1977, Distinctive features of some ancient river systems, in Miall, A. D., ed., Fluvial sedimentology: Canadian Society of Petroleum Geologists Memoir 5, p. 531-542.

Gill, J., and Gorton, M., 1973, A proposed geological and geochemical history of eastern Melanesia, in Coleman, P. J., ed., The Western Pacific Island Arcs, Marginal Seas, Geochemistry; New York, Crane Russak & Co., p. 543-566.

Goguel, J., 1948, Introduction a l'etude mecanique des deformations de l'ecore terrestre: Mem. Expl. Carte Geol. Fr., p. 530.

Hamilton, W., 1973, Tectonics of the Indonesia region: Geological Society of Malaysia Bulletin, v. 6, p. 3-10.

Hampton, M. A., 1975, Competence of fine-grained debris

flows: Journal of Sedimentary Petrology, v. 45, p. 834-844.

Harms, J. C., and Fahnstock, R. K., 1965, Stratification, bedforms and flow phenomena (with an example from the Rio Grande), in Middleton, G. V., ed., Primary sedimentary structures and their hydrodynamic interpretation: Society of Economic Paleontologists and Mineralogists Special Publication 12, p. 84-115.

Harrison, T., 1975, Tektites as "date markers" in Borneo and elsewhere: Asian Perspectives, v. 18, p. 60-63.

Hein, F. J., and Walker, R. G., 1977, Bar evolution and and development of stratification in the gravelly, braided, Kicking Horse River, British Columbia: Canadian Journal of Earth Sciences, v. 14, p. 562-570.

Hose, R. K., and Danes, Z. F., 1973, Late Mesozoic to early Cenozoic structures, eastern Great Basin, in deJong, K. A., and Scholten, R., eds., Gravity and tectonics: New York, John Wiley and Sons, p. 429-441.

Hubbert, M. K., and Rubey, W. W., 1959, Role of fluid pressure in mechanics of overthrust faulting. I. Mechanics of fluid filled porous solids and its application to overthrust faulting: Geological Society of American Bulletin, v. 70, p. 115-166.

Jackson, R. G., 1977, Preliminary evaluation of lithofacies models for meandering alluvial streams, in Miall, A. D., ed., Fluvial sedimentology: Canada Society of Petroleum Geologists Memoir 5, p. 543-576.

Karig, D. E., 1973, Plate convergence between the Philippines and the Ryukyu Islands: Marine Geology, v. 14, p. 153-168.

Kehle, R. O., 1970, Analysis at gravity sliding and orogenic translation: Geological Society of America Bulletin, v. 81, p. 1641-1664.

Kleinpell, R., 1954, Reconnaissance Geology and oil possibilities of Northern Luzon: Geological Report of the Philippine Oil Development Co., Manila, Philippines (unpublished).

Lajoie, J., 1979, Volcaniclastic rocks, in Walker, R. G., ed., Facies models: Geoscience Canada Reprint Series 1., p. 191-200.

Lemoine, M., 1973, About gravity gliding tectonics in the western Alps, in deJong, K. A., and Scholten, R., eds., Gravity and tectonics: New York, John Wiley, p. 201-216.

Lopez, S. M., 1971, Notes on the occurrence of fossil elephants and stegodonts in Solana, Cagayan, Northern tology of Liwan plain: Seminar in Southeast Asian Prehistory and Archaeology, Manila, Philippines (unpublished report).

Mammerickx, J., Fisher, R. L., Emmel, F. J., and Smith, S. M., 1976, Bathymetry of the East and Southeast Asian Seas: Geological Society of America Map and Chart Series, MC-17.

McGowan, J. H., and Garner, L. E., 1970, Physiographic features and stratification types of coarse-grained point bars: Modern and ancient examples: Sedimentology, v. 14, p. 77-111.

Miall, A. D., 1979, Deltas, in Walker, R. G., ed., Facies models: Geoscience Canada Reprint Series 1, 43-56.

Moody-Stuart, M., 1966, High and low sinuosity stream deposits with examples from the Devonian of Spitsbergen: Journal of Sedimentary Petrology, v. 36, p. 1102-1117.

Movius, H. L., 1949, The lower Paleolithic cultures of southern and eastern Asia: Transactions of the American Philosophical Society, v. 38, p. 329-420.

Mullineaux, D. R., and Crandell, D. R., 1962, Recent lahars from Mt. St. Helens: Geological Society of American Bulletin, v. 73, p. 855-870.

Murphy, R. W., 1973, The manila trench-west Taiwan foldbelt-flipped subduction zone: Geological Society of Malaysia Bulletin, v. 6, p. 27-42.

O'Keefe, J. A., 1963, Tektites: Chicago, Univ. of Chicago Press, 228 p.

Pierce, W. G., 1973, Principal features of the Heart Mountain fault and the mechanism problem, in deJong, K. A., and Scholten, R., eds., Gravity and tectonics: New York, John Wiley, p. 457-472.

Ragland, P. C., Stirewalt, G. L., and Newcomb, W. E., 1976, A chemical model for island arc volcanism, western Luzon, the Philippines: Geological Society of America Abstracts with Programs, v. 8, p. 1056.

Reineck, H. E., and Singh, I. B., 1975, Depositional sedimentary environments: New York, Springer-Verlag, 439 p.

Roedder, D., 1977, Philippine arc system--collision or flipped subduction zones?: Geology, v. 5, p. 203-206.

Rust, B. R., 1978, Depositional models for braided alluvium, in Miall, A. D., ed., Fluvial sedimentology: Canadian Society of Petroleum Geologists Memoir 5, p. 605-625.

Rust, B. R., 1979, Coarse alluvial deposits, in Walker, R. G., ed., Facies models: Geoscience Canada Reprint Series 1, p. 9-21.

Santos-Ynigo, L., 1964, Island arc feature of the Philippine archipelago: 22nd International Geological Congress, New Delhi, Proceedings Section 11, p. 369-386.

Sartono, S., 1973, On Pleistocene migration routes of vertebrate fauna in southeast Asia: Geological Society of Malaysia Bulletin, v. 6, p. 273-286.

Schmincke, H. V., 1967, Graded lahars in the type section of the Ellensburg Formation, south-central Washington: Journal of Sedimentary Petrology, v. 37, p. 438-448.

Smith, N. D., 1974, Sedimentology and bar formation in the upper Kicking Horse River, a braided outwash stream: Journal of Geology, v. 82, p. 205-223.

Sparks, R. S. J., 1976, Grain size variations in ignimbrites and implications for the transport of pyroclastic flows: Sedimentology, v. 23, p. 147-188.

Sparks, R. S. J., Self, S., and Walker, G. P. L., 1973, Products of ignimbrite eruptions: Geology, v. 1, p. 115-118.

Sparks, R. S. J., and Walker, G. P. L., 1973, The ground surge deposit: A third type of pyroclastic rock: Nature, Phys. Sci., v. 241, p. 62-64.

Sparks, R. S. J., and Walker, G. P. L., 1977, The significance of vitric air-fall ashes associated with crystal-enriched ignimbrites: Journal of Volcanology and Geothermal Research, v. 2, p. 329-341.

Steel, R. J., Maehle, S., Nilsen, H., Roe, S. L., and Spinnangr, A., 1977, Coarsening upward cycles in the alluvium of Hornelen Basin (Devonian), Norway: Sedimentary response to tectonic events: Geological Society of America Bulletin, v. 88, p. 1124-1134.

Tamesis, E. V., 1976, The Cagayan valley basin: A second exploration cycle is warranted: SEAPEX Program, Offshore Southeast Asia Conference Paper 14.

van Bemmelen, R. W., 1970, The geology of Indonesia, Vol. 1A 2nd edition.

Varnes, D. J., 1958, Landslide types and processes, in Eckel, E. B., ed., Landslides and engineering practice: National Research Council, Highway Research Board Special Report 29, p. 20-47.

Vergara, J. F. et al., 1959, Preliminary report on the geology of Eastern Cagayan Region: Philippine Geologist, v. 13, p. 44-55.

Verstappen, H., 1975, On paleo-climates and landform development in Malaysia: Modern Quaternary Research in Southeast Asia, v. 1, p. 3-35.

Visher, G. S., 1965, Fluvial processes as interpreted from ancient and recent fluvial deposits, in Middleton, G. V., ed., Primary sedimentary structures and their hydrodynamic interpretation: Society of Economic Paleontologists and Mineralogists Special Publication 12, p. 116-132.

von Koenigswald, G. H. R., 1956, Fossil man from the Philippines: National Research Council of the Philippines. Originally paper No. 22, Proceedings of the 4th Far Eastern Prehistory Congress.

von Koenigswald, G. H. R., 1958, Preliminary report on the newly-discovered stone age culture from northern Luzon, Philippine Islands: Asian Perspectives, v. 2, p. 69-70.

von Koenigswald, G. H. R., 1967, Tektite studies. IX. The origin of tektites: Proceedings of the Koninklijke Nederlandsche Akademie von Wettenschappen, v. 70, p. 104-112.

von Koenigswald, G. H. R., and Gosh, A. K., 1973, Stone implements from the Trinil beds of Sangiran, central Java: Proceedings of the Koninklijke Nederlandsche Akademie von Wettenschappen, v. 76, p. 1-34.

Walker, G. P. L., 1971, Grain-size characteristics of pyro-
    clastic deposits: Journal of Geology, v. 79, p. 696-714.

Walker, R. G., and Cant, D. S., 1979, Sandy fluvial systems,
    in Walker, R. G., ed., Facies models: Geoscience
    Canada Reprint Series 1, p. 23-31.

Wasson, R. J., and Cochrane, R. M., 1979, Geological and
    geomorphological perspectives on archaeological sites in
    the Cagayan Valley, Northern Luzon, the Philippines: A
    report to the National Museum of the Philippines
    (unpublished).

Williams, H., and McBirney, A. R., 1979, Volcanology, San
    Francisco, Ca., Freeman, Cooper and Co., p. 397.

Wright, L. D., 1978, River deltas, in Davis, R. A., ed.,
    Coastal sedimentary environments: New York, Springer-
    Verlag, p. 5-68.

DATE DUE